IT ALL BEGAN WITH DAISY

Sonia Jones

IT ALL BEGAN WITH DAISY

An Eleanor Friede Book

E. P. DUTTON · NEW YORK

Published in the United States by E. P. Dutton,
a division of NAL Penguin Inc.,
2 Park Avenue, New York, N.Y. 10016.

Published simultaneously in Canada by Fitzhenry & Whiteside Limited, Toronto.

Library of Congress Cataloging-in-Publication Data
Jones, Sonia.
     It all began with Daisy.
     "An Eleanor Friede book."
     1. Dairying—Nova Scotia—Lunenburg.   2. Peninsula
Farm (N.S.)   3. Jones. Sonia.   I. Title.
SF233.C2J66   1987   637'.09716'23   86-29201

ISBN 0-525-24540-5

COBE

Designed by Steven N. Stathakis

10 9 8 7 6 5 4 3 2 1

First Edition

This book is dedicated to Gordon, Valerie, and Vicki, with my love.

Contents

Acknowledgments

No company can succeed without the extraordinary efforts of its key personnel, so it is with deep gratitude that we thank the following current employees for their dedication to Peninsula Farm: Cheryl Allen, Eugene Barkhouse, Sylvia Booth, Chester Bowles, David Bruce, Isobel Burbidge, Margie Clark, Ralph Clark, St. Clair Corkum, Valerie Curwin, Patricia Duffy, Marilyn Emmerson, Richard Finney, Wanda Garnier, Pamela Geddes, Karen Hall, Jerry Johnson, Blair Landry, Bart Lausanne, Roger LeBlanc, Cheryl Lohnes, Andrew MacDonald, John MacDonald, Mark MacInnis, Kerry Mason, Susan Murray, James Pittman, Jr., Robert Pougnet, Mark Rafuse, Shirley Richardson, James Rutledge, Jr., and William Towndrow.

Although we were unable to mention in the text the names of all the chain-store executives, store managers, grocery managers, and dairy-case stockers who helped us along the way, we would like to extend general thanks to all those associated with the following companies: ADL Foods, Amalco Foods, Atlantic Wholesalers, Bolands, Capitol Stores, Central Dairies, Co-op Stores, Dominion

Stores, the Food Group, Sobeys, Wade's Ltd., F.E. Wade's, and Willett Foods. Our thanks, too, to the schools, hospitals, and institutions that carry our products.

It is impossible to list the many suppliers that have sold us top-quality ingredients over the years, but we would particularly like to single out Atlantic Wholesalers, Farmer's Cooperative Dairy, IMA Aquatic Farming, Henry Knol, Lantic Sugar, Mapleton Enterprises, Oxford Frozen Foods, and Westvale Foods.

We are grateful to the following people for giving us valuable professional advice: Wayne Fudge and Byron Wilson (Doane Raymond); Greg Silver, Denise Saulnier, and Al Godbout (Communications Design Group); Fred Kennedy, Neil Smith, and Peter Tye (Physical Distribution Advisory Service); Reg Cluney (McInnes, Cooper, and Robertson); Morton Amster (Amster, Rothstein, and Ebenstein); Marshall Margolis (Kitamura, Yates, and Margolis); and Martin Stuchiner.

We also thank the following brave souls for always coming when we needed them (or for working all night when necessary): Reg Bell (Liz-One Irving), Bruce Campbell (Lunenburg Electric), Mike Hull and Larry Oickle (L & B Electric), Rick Joudrey (Thermo King), Merrill McCarthy (Farmers' Cooperative Dairy), Leonard Stevens, Klaus Willner (Quality Containers), Al Young (Young's Refrigeration), and Leonard Zwicker (Len's Plumbing and Heating).

And to my good friends and neighbors—Gibbie and Artie Rhuland, Elvin and Peter Falkenham, Junior Nodding, and Harry Pernette—my deepest appreciation.

We would like to thank the following government agencies for their financial assistance at crucial moments during our growth: the Nova Scotia Department of Development, the Nova Scotia Department of Agriculture and Marketing, the Department of Regional Industrial

Expansion, the National Research Council, the Department of Employment and Immigration, and Agriculture Canada.

Hats off to the numerous, unsung, unknown customers who have been faithful to us over the years in spite of the quality variations that have occasionally resulted from our vigorous muddling!

And finally, special thanks to Eleanor Friede and Barbara Bowen for their invaluable help in guiding and developing this project in all its many phases.

Foreword

Many people have watched Peninsula Farm progress from a cow and kitchen stove operation to a computerized dairy plant with a network of tractor-trailers distributing yogurt to stores throughout the Atlantic provinces. Most of these people ask me the same question: How did you do it? It sounds like a straightforward query, one that could be answered by mentioning some of the characteristics of the typical entrepreneur (hardworking, tenacious, energetic, ambitious, innovative, persistent, flexible, and naïvely optimistic). A thoughtful answer, however, would have to put the question into some kind of context, and I was always far too busy milking cows and shoveling dung to tackle the subject.

But now that Sonia has written a book about Peninsula Farm, I have no place to hide. I can't go on pretending I'm too busy to take a look at the various strategies that led to our growth and ultimate success, and I can't avoid the issue by telling people to consult the business management experts, for every company constitutes a case study in its own right. What makes it unique is basically a question of heredity and environment, for a newborn company's personality is partly shaped by the philosophy of its creators and partly by the business climate prevailing at the time.

Peninsula Farm was fortunate to be conceived and nurtured in a small, rural community where business practices reflected the traditional values of generations of hardworking individuals accustomed to dealing shrewdly but honestly with one another. The owners of big corporations seemed sensitive to the needs of smaller companies and were not averse to lending a hand where necessary. The bottom line was not always a primary consideration with the decison makers, and this made it possible for us to muster our forces in the early years when we were particularly vulnerable. The provincial and federal governments were also helpful in the beginning when we needed financial assistance to boost our growth. The business environment was generally excellent for the development of a new company, although there were certain apparent drawbacks such as widespread consumer resistance to yogurt and high distribution costs because of the sparse population of Nova Scotia. But these factors could also be seen as benefits if examined in another light, for they forced us to grow slowly while we were still in our error-prone infancy.

The blood and guts of a company, however, and the major contributors to its success or failure are the people who create it: the entrepreneurs. This is not to say the employees are unimportant. In our own case, we concluded that attitude was the most decisive factor in personnel selection, and we tried hard over the years to choose a team whose members had the maturity to work well together and the insight to understand that the good of the company and their personal welfare were one and the same thing. But Sonia and I were the ones responsible for setting standards, deciding policy, and roughing out the long-term goals of Peninsula Farm. We knew that if we were to be successful we would have to keep a close eye on every detail of our operation, learning by our mistakes as we went along and never depending too much on outside advice. It came as something of a shock to us that we had to plow back all the profits for the first eight years, but this sacrifice certainly helped to put the business on a firm footing. Once we had come to terms with the fact that we had to be on call twenty hours a day for seven days a week, the rest was smooth sailing.

As a husband-and-wife team we divided up the labor between us according to our individual talents and interests without coming to blows, but it has not always been that easy. There was a time when I sincerely thought our marriage would not survive the honeymoon, for Sonia had the audacity to beat me at chess three times in a row. I sulked outrageously until she rescued our relationship by calmly explaining to me that her gifts were now mine and that together we possessed abilities that were synergistic. Our marriage was enriched by our differences, for it was after all a cooperative effort, not a competitive one.

I have never forgotten this lesson. Now I am free to celebrate her successes, and she mine. We bring two differing views to every problem and choose the best solution without reverting to rivalry or ego involvement. Peninsula Farm has flourished as a result of Sonia's wide-ranging talents, which are so natural to her she seems unaware of their existence. People often ask me how she manages to be an outstanding professor, author, entrepreneur, mother, and wife all at the same time without falling apart at the seams. The answer is simple: she is an exceptional woman, and as far as Peninsula Farm is concerned, she is the true cornerstone of our success.

But success is ephemeral, and there is never any guarantee that it will last forever. As I write this I don't know what the future holds for Peninsula Farm. Will our giant competitors, such as Beatrice, Kraft, and Yoplait, find a way to put us down? Will our chain-store customers "deal" us out of the marketplace? Will the distributing dairies buy our shelf space? Or will we become an international conglomerate, licensing dairies all over the world to make our products? What I do know is that we will start the new year as we finished the old—interested only in making the best yogurt, fighting always to be a successful business, and preserving our quality no matter what the cost.

Gordon Jones

Mauna Lani Bay Hotel
Island of Hawaii
December 1985
(Lest you think there are no rewards)

1

Exodus and Genesis

Good-bye to New York

To this day Gordon and I stand back after a hard day's work on the farm and survey our land with a mixture of awe, bewilderment, and disbelief. We are awed by the beauty of the green fields that jut out into a sparkling ocean and bewildered by the peacefully grazing cows who seem to take it perfectly for granted that they are safe in the hands of a couple of bumbling urbanites like ourselves. But what we find most difficult to believe is that all this is *real*, that we actually *live* here, that the timothy, the alfalfa, the spruce trees, the red barn, and the cozy white farmhouse all belong to us and to us alone. Twelve years ago we were snugly ensconced in our matchbox-size Manhattan apartment, unaware that we were soon to become the baffled owners of a tiny seaside farm on the south shore of Nova Scotia, a farm whose prosperity and ecological integrity would rest entirely in the care of two people whose combined knowledge of grass and animals consisted of a trip to the Bronx Zoo, three strolls in Central Park, and my own intimate acquaintance with the pigeons that make their home on the ledges of Harvard's Widener Library.

If our move to Nova Scotia can be explained by any one factor, the Widener pigeons deserve much of the credit. They taught me that animals are much saner than humans, and this was an important lesson for me to learn during my years as a graduate student. For one thing, they had the good sense to spend their days on the *outside* of Widener Library, whereas I spent half a decade laboring away deep inside the dark, sunless stacks, cramming my brain with facts whose relevance to the real world would have been considered rather tenuous by the pigeons. Unfortunately it never occurred to me to consult them about my objectives in life, so I doggedly pursued my studies until one day I found myself standing in front of Widener wearing a black gown and a mortarboard along with a number of other people in similar disguises. I emerged from the ceremony bearing a parchment declaring me to be a bona fide doctor of philosophy, and the next day I joined my husband in New York, fully expecting to take Manhattan's academic community by storm.

I soon learned, however, that academics are not easily impressed, especially by neophyte scholars. One department chairman raised a nostril in delicate disdain at the sight of the skimpy list of publications on my curriculum vitae, and another informed me, rather callously I thought, that New York was literally teeming with applicants far more experienced than I was.

"Don't let them get you down," said Gordon, who patiently listened to my lachrymose tales of woe every evening when I came home bloodied but pretending to be unbowed. "That's what those professors are trained for, to be critics. And besides, they're not going to look favorably on anyone but themselves, so don't take it personally."

"But what am I supposed to *do?*" I wailed, hoping

to make my predicament seem a little more dramatic by gnashing my teeth, but I wasn't sure how to go about it.

"Just keep on going to those interviews. Something will turn up, you'll see."

Gordon is the one person in this world who would never even consider quitting. Adversity galvanizes him into action, and he meets challenges with the kind of unflagging zest that marks the true entrepreneur. As a result he had built a thriving management consulting firm from an idea that had originally provoked nothing but good-natured sneers from potential partners.

"Computers? You've got to be kidding," they said in chorus. "Computers will never go anywhere. They're too expensive, and they take up too much room. Only the government and the military and the banks can afford the damn things. And you can be sure as hell they're not going to need *you* for any consulting. C'mon, Gordon, get back down to earth!"

But Gordon was not interested in being earthbound, so he struck out on his own and quietly opened a one-man consulting firm in a back room near Union Square. With the passing years tubes gave way to transistors and finally to solid-state technology, and Gordon rode the growth curve. As corporations and smaller companies began to acquire computers, Gordon and his burgeoning team of consultants were right on the spot to advise them about programming and systems analysis. The back room at Union Square was eventually replaced by a suite of offices on Park Avenue, and it wasn't long before his old cronies, the potential partners of yesteryear, turned up to find out how he was progressing.

"Not bad, not bad," they breathed, looking the place over with narrowed eyes. "You've come a long way in fifteen years. You've been lucky, ol' buddy."

"Lucky?" I echoed indignantly after they left. "Don't they realize how hard you worked? How can they say that? What you've done has nothing to do with luck!"

"Oh, you'd be surprised," he grinned. "Some of the best business decisions are purely a matter of chance. But still, they're decisions, and right or wrong you build on them and keep going. At least that's how it's been for me, anyway. If one thing doesn't work out you take another tack, and if that's no good you try something else, and eventually it all comes together. It's a question of muddling through, really. But it's hard work, of course. I guess I'd have to call it *vigorous* muddling."

Gordon's theory of vigorous muddling was soon applied to my own rather disappointing performance in the job market. By midsummer it had become abundantly clear that my Harvard Ph.D. was not going to land me a position at a university in or near New York City, so one morning Gordon rather laconically suggested that I broaden my horizons to include the rest of North America.

"All you have to do is roll up your sleeves and write to every department where there's a job opening advertised for next year. You can't miss. I'll have my secretary feed your cover letter through the electronic typewriter, and she'll have twenty originals for you in no time. Then you can sit down and sign them, and she'll roll them through the stamp machine and have them in the mail the same day. I guarantee you'll get some action."

"But Gordon," I said, wondering just how seriously to take him, "what if I get an offer from some place far away, like the University of Hawaii or something like that?"

"Terrific!" he cried with unbridled enthusiasm. "I've always wanted to live in Hawaii. Great place for sailing. Even better for surfing. We'd have to live away from Honolulu, though. Too many people. You wouldn't mind commuting, would you?"

"But what about your business?" I said, trying not

to sound too alarmed. "How can you even *think* about walking away from your business? What will you do with it?"

"I'd sell it, of course. What else would I do with a business? I can't very well pack it up in a suitcase and take it with me."

"But Gordon, your business is flourishing. You're a success. Why on earth would you want to *sell* it?"

"Why? Because there's nothing new about it now. I've already taken it as far as it's going to go. We're not on the frontier anymore. Everybody's into computers nowadays, and it's getting too serious. It's getting to be a drag, as a matter of fact. It used to be a lot of fun in the beginning, but now the challenge is gone. All I do is sit in my nice air-conditioned office and I pick up the phone and I talk to some jaboney for a while and when I hang up I know I've made some banker richer, but I keep asking myself what the hell it's all about. I mean, when do I get to see the light of day? I'm always in such a rush to help more people make more money, but I don't even know what I'm doing it for. Rush, rush, rush, that's the story of my life. You know what happened to me the other day? I was in the men's room when I suddenly realized I was standing there trying to *pee* in a hurry, for God's sake!"

"Gordon," I said, making an effort to sound calm and reasonable, "I can understand how you feel. Really I can. So maybe what you need, what we both need, is a nice long vacation. A cruise or something. But to give up your *business* . . ."

"Look at me," he said, rather impatiently. "Just look at me for a minute. What do you see? Go ahead, tell me."

"What do you mean?" I asked suspiciously.

"I'm forty-two years old, Sonia. Do you know how that feels? No, how could you? You're ten years younger than I am."

"Eight years younger."

"I've lived in New York for forty-two years," he went on, dismissing my correction with a wave of the hand. "Forty-two years!"

And so it was that we embarked on a sketchy plan that was eventually to lead to some very vigorous muddling indeed. The idea was for me to apply to any university that appealed to me, with the strict understanding that the campus be near an ocean so that Gordon could indulge his lifelong passion for sailing. He was sick to death of trying to maneuver his ketch through the weekend traffic jams on Long Island Sound. All he needed in life, he said dreamily, was a little piece of ocean where he could have some real elbow room, and a bit of land where he could build and launch a new yacht with his own hands. This ambition, he felt, was a modest one, and since it included a strong desire to father and help care for a baby or two, I found it enormously attractive. We had wanted to start a family for a number of years, but graduate studies and business pressures had forced us to postpone it for longer than we had intended. Now it seemed that a plan was emerging that would allow us both to satisfy our personal and mutual yearnings, and as the plan grew, so did our excitement.

"Just think, we'll be able to fill our lungs with fresh air all day long," said Gordon happily. "No more coughs, no more stinging eyes."

"We'll have some land of our own, and we'll grow our own food, and I'll make all our meals from scratch."

Gordon's eyes lit up. "I'll keep as many dogs as I like, and I'll never have to worry about a pooper-scooper."

"We'll be living among honest, hardworking people who would never *think* of mugging us or robbing us. We could *relax!*"

"Do you realize," Gordon went on, "that I've spent my whole life surrounded by concrete and asphalt? I'll be able to roll in the grass, and watch the leaves turn, and take walks in the woods, and feel the ocean spray on my face."

When Gordon's friends heard about our decision to turn our backs on the Big Apple and move to a paradise by the sea, the hoots of derision could be heard from coast to coast. They were more or less the same cronies who had spoken with such great authority on the bleak future of the computer, and now once again they were ready to regale us with gratuitous advice.

"Wake up, Gordon, can't you see what's going to happen? Sonia will be the breadwinner and everybody'll say you're tied to her apron strings. You'll be stuck in the house changing diapers while she's out hobnobbing with her colleagues. You're not going to like switching roles that way. You're not the type."

"Not the type, am I?" said Gordon, with an enigmatic smile.

"We give you one year, one and a half tops, and you'll come crawling back to New York with your tail between your legs, and *then* what'll you do? Start all over again? C'mon, Gordon, smarten up! You've got it made. What do you want to go and throw it all away for? If it works, don't change it."

"I'll try to remember that," said Gordon tactfully.

Nova Scotia bound

By January of the following year I had amassed *three* job offers from universities-by-the-sea. Two of them were from familiar areas of the United States coastline, but the one that really captured our imaginations came from a fascinating, faraway spot called Nova Scotia. The very sound of the name caused us to drift into peaceful reveries of rocky shores and endless pine forests, but we took it for granted that a trip to Dalhousie's campus would quickly dispel our romantic notions. We were wrong. It turned out that Nova Scotia did indeed have a plethora of rocky shores and evergreens, and they looked almost exactly as they did in our dreams. But what pleased us most of all was that Dalhousie (we later learned that it rhymed with frowsy, not floozy) was located in Halifax, a small city that offered all the amenities of a larger population area, but without the inevitable crowding. There was an excellent theater downtown, the university had a decent library, and best of all there was a modern cultural center on campus that was reputed to attract many of the best-known performers in North America and Europe. We decided then and there that Nova Scotia would become our new home.

We arrived in late spring, bringing with us only what we needed until we could find a permanent place to stay. This included three suitcases, a carton of books, the proceeds from the sale of Gordon's business, a case of Pampers, and a newborn baby girl, not necessarily in that order of importance. We set out at once to comb the coastline for a piece of land with enough water frontage to allow Gordon to harbor his eventual yacht. As for me, I was keeping a weather eye out for a cottage with a deck where I might spend the cool evenings nursing Valerie in a rocking chair and thinking great thoughts as I contemplated the distant horizon. My classes at Dalhousie, I mused with smug self-satisfaction, would directly reflect the inspiration that came to me on the waves that broke eternally over a glistening beach.

We ended up in the company of a youthful real estate agent who was blessed with a sense of hopeless, unquestioning optimism about what his company could do for us. Gordon put Valerie in a backpack and we valiantly traipsed along behind this dreadnought, who proceeded to drag us over marshland rife with black flies, through mosquito-infested woods abutting on two or three feet of shoreline, and over weed-clogged properties that were virtually inaccessible to anything but an army helicopter. After a few weeks our hopes and our patience began to flag, but it was Valerie who finally put her foot down. One morning she simply refused to allow herself to be dropped into her backpack, and that was the end of our fruitless treks along Nova Scotia's wild coastline. We had to admit that it all looked much better on a map.

But our friend the dreadnought was not one to give up easily. Shortly after Valerie's triumph over the back-pack, he was on the phone describing in ecstatic terms a small farm in Lunenburg, which he claimed was *just* what we were looking for. I knew right away that a farm was most definitely *not* what we were looking for and that Lunenburg was much too far away from Halifax to suit me, but Gordon couldn't resist a quick little drive to the country.

"We'll put Valerie in her car seat and she'll fall asleep right away," he said cajolingly. "And you'll get to see Lunenburg. They built a famous schooner there, the *Bluenose* they called it, but they had to build another one after it sank somewhere down in the Caribbean. Lunenburg is the saltiest little town you'll ever see, full of fishermen and lobster traps and scallop draggers and trawlers and boat builders . . ."

As soon as he mentioned boat builders I knew that our trip to Lunenburg was as inevitable as the tides, but what I didn't realize at the time was that Lunenburg was also going to be our destiny in more ways than one. This became clear to both of us the moment we laid eyes on the farm. It lay stretched out along half a mile of ocean frontage, green and inviting against the cool water. A dirt road connected the ample barn with the saltbox farm-house, and weather-beaten fences meandered along the property lines and off into the distant woods.

"This is it," said Gordon, filling his lungs with the salt air. "We're home."

"But Gordon," I protested, knowing very well I could never hope to persuade him to reconsider. "Aren't you being a little hasty? Are you sure you really want to get into farming?"

"Who said anything about getting into farming?" he said, putting his hand reassuringly on my shoulder. "Where is it written that you automatically have to be a full-fledged farmer just because you own a farm?"

"But isn't it kind of a shame to have a farm and not use it? What's going to happen to agriculture in Nova Scotia if farmers sell their land for real estate? What will people do for food if the farms go out of production?"

"I wouldn't worry about that. We're a long way from starving in Nova Scotia. And besides, we only have twenty-five acres here. Twenty-five acres more or less isn't going to make a drop of difference to the provincial economy."

"I don't know," I said dubiously.

"Listen, all I want to do now is relax and mind my own business. I want to sit back and drink in all this beauty every single day, till the skyscrapers and the telephones and the subways and the traffic jams are all washed out of my system. I don't want to know about agricultural economics. I just want to build my yacht in peace and quiet, and at the end of the day I want to stare at the sky and the ocean and the sea gulls. See them down there on that back field? Just look at them! *They* have the right idea."

As I contemplated the happily foraging sea gulls, I felt my apprehension slip slowly away. My thoughts rested nostalgically on the pigeons of Widener Library, who, like their brethren the sea gulls, had no difficulty whatsoever sorting out their priorities. Maybe Gordon was right. Life comes in phases, and it was time now for us to exchange the absurd hustle and bustle of our city days for a more sensible rural pace. We would live an idyllic life far from the madding crowd, and Gordon, who was obviously ready for a change, would have ample opportunity to develop new interests. I knew I could count on him to land on his feet, even if it meant spending extra time on vigorous muddling.

2

Lesson One

Surviving without the super

Every morning I would wake up thinking I was back in our own bedroom in New York, and yet something was wrong. A subliminal awareness of total silence would stubbornly nag me into reluctant consciousness. No honking, no air brakes, no clanking of garbage cans. I would open my eyes in a panic, wondering what had happened to the outside world, and then it would very slowly dawn on me that we had moved to a farm in Nova Scotia. That was when the *real* panic set in. Where was Lenny the superintendent if (God forbid) something should break down? Could we call him for advice in case of emergency? I already knew what his answer would be.

We were doomed to a life without Lenny, and I knew we would have to get used to it as quickly as possible. But what I didn't know, in those early, innocent, halcyon days on the farm, was that major catastrophes occur with distressing regularity in the country.

Our first one struck when we least expected it. We stumbled out of bed at the crack of dawn one morning and proceeded to the bathroom to perform our daily ablutions. No water.

"Oh God," I moaned. "What do we do *now?*"

"Don't worry," said Gordon cheerfully. "I'll take care of it. I'll go check the pump down in the barn and see what I can find out. I'll soon have it going again. I hope."

Lenny, I thought, would have been much less sanguine about the situation. He would have taken it as a personal insult that the pump had chosen to break down before breakfast, and on a *Sunday* yet! But Gordon set off for the barn with the buoyant gait and high spirits of a man unaccustomed to the endless tedium of leaky faucets, cracked plaster, and faulty wiring that Lenny found so patently uninspiring. Gordon had never had the opportunity of tackling anything quite so prosaic as a household water pump before, so he was not only eager to meet the new challenge but delighted to think that a victory over the pump would represent a giant step toward self-sufficiency, a concept that was beginning to appeal to him more and more every day.

As for me, I was content just to redecorate the dreary and rather tasteless farmhouse. My early morning anxiety would fade with the growing light of day, and it was never long before I was happily at work sanding banisters, varnishing trim, hanging wallpaper, laying carpet, piecing together tiles, or slapping creamy paint over the sickly green walls. Valerie was still too young, at the age of four months, to take a very active interest in either the purpose or the outcome of my labor, but I was glad for the company she provided. I was also grateful to her for vehemently insisting, sometimes at the top of her lungs, that I stop working at frequent intervals to relax with her on the rocking chair while she hungrily claimed her daughterly rights. It was during one of these quiet moments that Gordon came back from the barn, looking puzzled and frustrated.

"The pump didn't want to shut itself off, so there has to be a leak somewhere," he said, as he made his way down the cellar stairs. He clanged and banged around for a few minutes and then emerged from the stairwell brandishing a shovel.

"This is going to be hell," he declared. "I can't find the leak anywhere. I've checked all through the barn and the basement, so it has to be underground. I'm going to have to dig up the pipes."

It was a Herculean task, but Gordon was determined to do it himself, claiming that a backhoe would probably break the pipes in ten different places and make an un-sightly mess as well. He calculated that he could dig two feet of trench per hour, so with sixty feet of underground pipe the job would take him no more than three days at the most. He hadn't counted on the fact that Nova Scotian soil is made of an amalgam of clay and rock hard enough to foil the efforts of even the most determined digger, nor had he foreseen the toll his sedentary life had taken on his long back muscles, yet in spite of everything he kept on going until the sun went down. The clock was just striking ten when I heard his step on the back porch.

"I found it!" he cried, bursting through the kitchen door in a state of disheveled triumph. "I found the leak! I only had to dig about six feet of trench, and there it was. What luck!"

"Well, sit down, then, and take off your boots," I said, worried he might be on the verge of collapse. He paid no attention at all.

"There was a huge, gaping hole in the pipe as big as a *quarter*," he said, indicating with thumb and fore-finger the exact magnitude of our problem. "No wonder we had no water. You should have seen it! Thank God for the midnight sun, or whatever it is we have up here."

The next morning Gordon woke up in a chair-shaped position. At first we were both convinced that he was destined to stay chair-shaped forever, but after about half an hour of massaging I was able to coax his joints and muscles to loosen their grip. This they did with audible creaks and groans, punctuated by an occasional yelp from Gordon himself.

"Take it easy!" he protested, as I straightened out his left shoulder. He was on his feet again by that time, hobbling tentatively toward the stairs.

"*Now* where are you going?" I asked nervously.

"To the hardware store, to get some fittings. I won't be long. I'll bring you back some bottled spring water and then we can have a whole pot of coffee. I'm dying for some Mocha Java."

We were indeed beginning to feel the effects of our unexpected deprivation. The dishes were piled up in the sink, the toilet was developing a noticeably pungent odor, and Gordon, after his strenuous day in the trenches, was feeling distinctly uncomfortable about not being able to shave and shower. Valerie, on the other hand, was perfectly content with her world. She continued her customary cycle of nursing, sleeping, exploring, and soiling her diapers as though nothing unusual were happening. For a while it seemed that the problem of cleaning her was an insurmountable one without water, until it occurred to me to anoint her bottom with my best cold-pressed virgin olive oil. Valerie, unimpressed by the price tag of thirty-nine cents an ounce, busily went on presenting me with new opportunities for a fresh wipe.

"I did it!" yelled Gordon, appearing suddenly at the door. "I fixed the pipe! It was a masterly plumbing job, a true work of art. Now all you have to do is turn on that kitchen tap and voilà!"

I dashed to the sink with the greatest of expectations, but when I turned on the tap there was no voilà at all, only the hollow sound of some distant gurglings followed by a profound silence.

"I can't understand it," said Gordon, looking crestfallen. "I did everything the book said. I cut the pipe evenly and put a coupling between the two ends and tightened down the clamps, so why don't we have water?"

"Is the pump working?" I asked, struggling hard to suppress a frantic urge to make a long-distance call to Lenny in New York.

"Yes, I checked it out before I came back to the house. It's in prime condition. It looks just about new. It couldn't be the pump . . . *Prime* condition! Prime the pump! That's it! The whole system has to be primed. It can't draw. We have airlocks in the system, that's what the trouble is!"

Hay ain't hay

If it hadn't been for the airlocks in our water pipes, it might have been a very long time before we met our neighbor from the adjoining farm. He was a lean, hatchet-faced individual named Travis Oickle. He had heard, via Lunenburg's lively grapevine, that a professor had moved in next door, and he had modestly concluded that a professor was too lofty a personage for him to approach. But when Gordon assured him that the professor was only me and that we were desperately in need of water to prime our pump, then the reticent Travis burst miraculously out of his shell and came trudging to our rescue bearing buckets of fresh water from his well. It wasn't long before endless streams of cool, clear water came gushing from our taps, spilling over the crusty assortment of pots and dishes that had accumulated in the kitchen sink. Gordon broke out a bottle of beer for Travis while I joyfully busied myself with the creation of an enormous mound of deliciously warm soapsuds. Even Valerie seemed to understand that there was cause for celebration, for she made several sporting attempts to leap out of my arms and into the inviting water in the sink.

"I don't mean to be askin' another man his business," said Travis a little later as we sat around the table, "but what's you and the missus plannin' to do with this place, if you don't mind me askin'?" He leaned back in his chair and scrutinized us carefully.

Gordon thought that because we were Americans Travis was afraid we would soon be putting up hot dog stands along the shore, so he hastened to reassure him that we fully intended to leave the farm exactly as it was. He waxed very poetic in his attempt to convince Travis that we appreciated the bucolic scenery and would never dream of interfering with it in any way. Travis listened patiently until Gordon had talked himself out.

"Yes," said Travis, when Gordon paused to rest. "That's all very well and good, and it ain't my place to tell a man what to do, but I never seen nobody cut twenty-five acres of grass with a lawnmower before."

Gordon looked at him blankly. "So what do you suggest I do?"

"I ain't sayin' nothin'," said Travis quickly. "I never put my nose where it don't belong, but the farmers around here, they all has their own tractors and equipment for to make their hay."

"Well, can't somebody make mine too while he's at it? I'd pay him, of course."

"I ain't sayin' yes, and I ain't sayin' no. But a man has to make hay while the sun shines, like they say, and up here we ain't got much of that. It's every man for himself when the sun is on the grass, and I can't say as anybody around here will be wantin' to make your hay for you while his is layin' and gettin' moldy on the land."

"I don't know the first thing about making hay," said Gordon doubtfully.

"Ain't nothin' to it," said Travis, looking affectionately at his third bottle of beer. "I could take you o'er to the Massey-Ferguson place myself if you wants to know what to buy, and I'd show you everything you has to know about gettin' that there grass off the land. But you better think about movin' along pretty quick because next month comin' up is July already, and your grass ain't gettin' no shorter."

The next day Gordon and Valerie and I went to the Massey-Ferguson dealer just to look at the farm machinery and to get a notion of what it might cost to own the necessary equipment for making hay. When we left the Massey-Ferguson dealer we were the startled owners of a brand-new tractor complete with front loader, rotary mower, a baler, and a combination rake and tedder, all in matching red.

"Are you *sure* we know what we're doing?" I asked Gordon in a thin and quavery voice.

"I'm never sure of anything," he said happily. "But after listening to what the salesman had to say, it seems pretty obvious that the tractor and front loader will pay for themselves ten times over. We can use them for *moving* things. And did you see the wheels on the tractor? That thing will take you *anywhere*."

"I'm not sure I want to go, not at that price," I muttered.

For the next two or three weeks Gordon spent most of his waking hours making hay with Travis. The plan was for Travis to teach Gordon everything he knew about hay in return for the use of our equipment, for his own machinery, dating back to a time just prior to World War II, had grown singularly unreliable over the years.

Gordon's respect for Travis and his wide knowledge of meteorology and machinery increased with every pass-

ing day. Travis taught him how to ignore weather reports
and sniff the wind when deciding when to cut down a
field; he showed him how to rake the hay into tight rows
to keep it as dry as possible on rainy days; and he
explained how to change the blades on his mower and
the teeth on his tedder when they were damaged by
the omnipresent rocks that lurked malevolently in the
grass.

Gordon grew tanned and skillful. He threw himself
into his new project with the enthusiasm of a man too
long confined to his desk, and soon he was handling the
tractor with confidence and authority. Travis watched
approvingly from the sidelines, arms folded and legs apart,
proud to be in charge of the practical education of this
apparently intelligent but oddly ignorant city boy.

"We need you, Sonia," Gordon sang out one sunny
morning as I was putting away the last of the breakfast
dishes. "We've got the hay all baled now, and we need
you to push the bales into piles so I can throw them on
the hay wagon as Travis goes by with the tractor."

It was a perfect day for working in the fields. A light
sea breeze was gently blowing my hair as I approached
the hay bales that lay in neat parcels under a cloudless
sky. I welcomed this opportunity to enjoy some healthy
outdoor activity, and as I wrapped my fingers around
the two strands of twine on the first bale, I pictured
myself hoisting it over my head and walking elegantly
to the next bale, where I would gracefully lower it to the
ground. The bale refused to budge. It didn't take me long
to realize that a bale of hay weighs close to fifty or sixty
pounds, and it became abundantly clear to me that if I
didn't go back to the house for some gloves I would probably have no fingers or palms left by the end of the day.
I looked over the field in despair. It was as if a low-flying
airplane had circled our farm and gleefully dumped its

entire cargo of passengers' suitcases at five-foot intervals in every direction. If it hadn't been for Valerie demanding attention and sustenance at convenient times during the day, I doubt very much that I could have survived until evening.

"Now that's what I call an honest day's work," said Gordon as we got ready for bed that night. "It makes me feel good, you know. Every day I'm learning something more about self-sufficiency. It's about time we discovered how to stand on our own two feet. . . ."

I never heard the end of his monologue. My own two feet had long since given out, and I had fallen into an exhausted sleep.

Everybody has a beef

"You're going to have to get some critters to gnaw off the land," said Travis one morning in his usual laconic tone of voice.

"Critters?" said Gordon, not knowing what Travis was talking about. "Critters to gnaw off the land?"

"Now it's not up to me to tell a man what to do," said Travis predictably. "But my family cleared an' worked that there land for three generations to get it to look like it does today, and I sure wouldn't want to see it go back to alder bush. You take an' neglect that land and in two or three years it won't be no good for nothin'. It'd be a sin and a shame for you to do somethin' like that to your land. But I ain't sayin' nothin'. Don't get me wrong."

All this came as quite a shock to Gordon, who had always been of the opinion that grass was grass and never needed much tending other than an occasional mowing. But here was Travis, calmly telling him that no, it's not enough to mow the grass every summer, it has to be gnawed by critters, and what's more the critters must provide fertilizer for it on a year-round basis in order to prevent this troublesome grass from "running out."

"Okay then," said Gordon suddenly. "Let's go to the next auction and get some cattle. If I can turn a profit raising a few head of beef, I might just be able to amortize that farm equipment a little faster."

"Well," said Travis, casually sucking on his pipe. "You can't go to no auction till you pound them fence stakes down. They heaves up on account of the frost in the winter. Then there's the water fences got to be made. Them critters'll walk clear down onto the beach at low tide and in three minutes they'll be halfway to Lunenburg unless you build fences out into the water so they can't get around 'em. Won't take you more'n a couple of days."

But Travis had grossly miscalculated our level of incompetence. He had never met any klutzes from New York City before, so he had no way of predicting that we would quickly become entangled in several miles of barbed wire for starters. Then there was the question of the fence mallet. Travis brought his own mallet with him one morning and very kindly demonstrated to Gordon how to swing it up over his head and pound a fence stake into the ground. It took Travis about three whacks to drive the stake almost a foot into the earth, then he handed it over to Gordon. To Gordon's astonishment the mallet seemed to be magnetized to the ground.

"God in heaven, Travis, what on earth is this thing made of, anyway? It must be a hundred percent *lead!*"

"Just swing it up o'er your head," grinned Travis. "Don't do no good to lift it slow. You'll get a hernia that way."

Gordon took his advice and heaved it into the air, then promptly fell over backward as his arms followed the arc of the mallet. Travis guffawed.

"Don't seem heavy to me," he remarked, lifting the mallet effortlessly over his head and slamming it down on another fence stake. The whole earth trembled.

"Give me that thing," said Gordon, taking the mallet from Travis. He raised it above his head and hit the fence post with such force that his feet rose six inches off the ground, but the post remained impervious to the blow. He hit it again, this time redoubling his efforts, but the recalcitrant post refused to move by even so much as a fraction of an inch. He turned and looked sheepishly at Travis.

"I think there must be a rock or something under that post. I can't get it to go down."

"Here, I'll give it a try," said Travis magnanimously.

Thwock, and the post plunged four inches into the earth. *Thud*, and it sank another four. I could feel the vibrations in the pit of my stomach.

By the time Gordon had fully prepared the farm to receive the long-awaited critters, summer was well under way and our clumpy, green meadows were beginning to look as though they could do with some energetic gnawing. When auction day finally arrived, we all bundled into the cab of Travis's cattle truck and soon we were rattling and grinding our way up to the Annapolis Valley, where the province's finest beef cattle were said to dwell. It wasn't long before we were comfortably settled in an outdoor arena watching a parade of cattle troop by, while the auctioneer kept up a nonstop volley of totally unintelligible prattle, punctuated occasionally with the English word *sold*, which he bellowed triumphantly into a microphone. Valerie took a lively interest in the various creatures on display, and lost no opportunity to answer their every call with a loud *moo* of her own.

Travis helped us become the proud owners of a grand

total of eleven beef critters. Five of them were cows, meaning that they had borne more than one calf, and the other six were heifers, a name used to describe virginal critters or the mothers of only one calf. As for their breed, one was a Charolais, two were Black Angus, and the rest of them were Herefords, a popular breed that came highly recommended by Travis because they were reputed not only to have tender meat but to be relatively easy to handle, even by the likes of us.

"Well, we've got ourselves a regular working farm," said Gordon that evening as we stood watching the eleven critters happily gnawing off the land. They had stumbled down the ramp of the truck and gone straight to work doing their critterly duties.

"I hope it doesn't cost too much to feed them when they're in the barn over the winter," I said with some trepidation.

"What can it cost?" said Gordon confidently. "We've already got all the hay they can eat, so that won't cost us anything. And anyway, Travis said it himself. This farm needs a herd of cattle to keep the land in shape, and that's just what we have now. We went and got ourselves eleven lawn mowers, and I'd say they're doing a pretty good job."

I was struck by how completely relaxed and contented the critters seemed to be in their new surroundings. Not one of them showed the slightest sign of being upset, depressed, or homesick. They stood in comfortable little groups, chewing their cud and aimlessly swishing at flies. One thing appeared to be absolutely certain: the cost of the tractor and farm equipment that had made the hay that lay in the loft concerned the critters not at all.

3

Geometric Progression

The real thing

By late November our beef critters were all comfortably settled in the barn for the winter. Gordon had correctly predicted that they would cost us nothing more in terms of cold cash now that the farm machinery was paid for, but we had both grossly miscalculated their cost in pure, unadulterated labor. To begin with, the interior design of our barn left much to be desired. It soon became apparent to us that the floor plan had been created by some anonymous architect who must have borne a secret grudge against farmers, for it had obviously been his intention to make certain that anyone working in this barn would do maximum penance. As a first step, he had decided that the water bowls should all be nailed to the outside walls. By this means he was able to accomplish several steps in the development of his medieval torture program. The sills were always wet, thus ensuring that the wood around the bottom edge of the barn would be in a state of perpetual rot. This not only provided convenient holes for the comings and goings of mice and other assorted wild creatures, but it also allowed the frigid winter air to enter and freeze the very water bowls that were causing the problem in the first place. Not only that, but we both lived in a state of constant fear that a lusty sneeze from one of the critters would bring the whole barn down upon our heads.

As if this were not enough, we soon discovered that the absence of an aisle in front of the critters made it very difficult for us to feed them. We were forced to approach them from behind while balancing a bale of hay on our shoulders; then we had to shove their massive bulks aside so that we could reach the mangers. This turned out to be a game that provided the cattle with their daily entertainment, for as we pushed on their hips and rib cages to move them over, they took infinite delight in pushing back just hard enough to make themselves utterly unbudgeable. When we finally did manage to dump some hay in front of their heads, it would eventually get soaked with the water they sloshed over from their ill-placed water bowls, causing enough wastage to make a noticeable difference in our feed budget.

Our worst problem turned out to be the positioning of the manure troughs. The barn had originally been designed for Holstein cows, which are about a foot longer than Herefords and other beef cattle. Consequently our critters were dropping their manure well short of the troughs and were then proceeding to wallow in it. The whitewashed walls were soon colored a lovely mud brown in an arc described by the longest tails, and we were faced with the daily task of not only shoveling manure from the stalls to the trough, but also scraping almost equal quantities from the walls and from the hides of the uncomplaining critters. We deposited the results of our labor into two wheelbarrows that we then pushed along a plank leading to the manure pile behind the barn. At first we found this chore to be reasonably manageable and even somewhat exhilarating, for we were quick to admit that our barn work was providing us with the healthy exercise we lacked when we lived in the city. But as winter drew on, we began to notice that the plank leading up the manure pile was getting steeper every day. Being from New York we were used to shoveling it, but here we were shoveling the real thing, and I, for one, was finding the work too slippery for my limited abilities. It was during an episode of howling wind and freezing rain that I at last had to confess to Gordon that I could no longer walk the plank with my wheelbarrow.

"It's no use, Gordon," I sighed, my eyes stinging with the icy wind. "I'm going to have to let you do the rest. The plank is covered with a sheet of ice, and it's up on a forty-five-degree angle. What I need is a pair of hobnail boots!"

"Just leave it to me," said Gordon cheerfully. "It's about time you gave yourself a rest. You shouldn't be doing all this heavy work anyway, not when you're pregnant."

He gave me a quick, sideward glance as he passed by with his wheelbarrow, wondering, no doubt, if I was going to give him a lecture again about Chinese women working in rice paddies right up until the moment they gave birth. I thought better of it. The shrieking wind had almost blown me off the manure pile, and I wasn't eager to take a fall.

"I'll say one thing, though," said Gordon as he came slipping and sliding back into the barn with his empty wheelbarrow. "Our manure pile has got to be the highest point in Lunenburg County. Have you noticed the view from up there? On a clear night you can see forty miles in every direction."

"Get away," I said, taking a swipe at him with my muck brush.

"No, I'm serious," he said, leaning on his shovel and smiling. "Just think about it for a minute. We've been working in this barn almost all winter, right? Okay, so did you know that the average cow produces her own weight in manure every three weeks? No, I'm not kidding! I read it in *The Dairy Breeder*. So figure it out. We have eleven critters and we've been working about, what, about twelve weeks, right? So that's forty-four times the weight of a cow, which is roughly half a ton, let's say. So you're looking at a twenty-two-ton manure pile out there!"

"No wonder we've got calluses," I said, peering out at the manure pile with a growing feeling of accomplishment. "And to think we did it all one shovelful at a time!"

Once back in the house, however, my self-congratulatory mood began to evaporate. My father would turn over in his grave, I thought, if he could ever know I was using my expensive education to create a twenty-two-ton manure pile.

"Reach for the stars, my darling daughter," he used to intone in his deep, stentorian voice. "You have only one life to live, so you must make every moment count. Fear nothing, stop at nothing, and always remember that your best is never good enough."

He would speak slowly, enunciating every word with his perfect British accent so that each phrase seemed weighty with significance. I could sense that his tongue was planted firmly in his cheek, and yet at the time I knew he was completely serious about everything he said.

"The world is an unsubstantial sort of place, a place of vanity and fleeting desire. You can lose everything you own in the blink of an eye. But there's one thing they'll never be able to take away from you, and that's your brain. Develop your brain in every waking moment. Make your life a great intellectual challenge, and let your riches be those of the heart and the mind. Collect experiences, seek wisdom, and above all remember always that the whole world is nothing but one vast desert of rampant mediocrity. Rise above it, my darling daughter, no matter what the cost, and reach for the eternal stars."

Unfortunately my father had never been one to conform too closely to what he preached. He was every bit an entrepreneur, and his riches were altogether too much of this world. He had worked all his life to help establish a small empire called Technicolor Films, Inc., but ultimately they came and took it away, just as he had always feared. Technicolor fell victim to a stock market raid in the early 1960s, and my father never really recovered from the blow. He tried to organize a proxy fight but he died of a heart attack before he could get anything under way. He was greatly missed by his wife and six children, and all of us, I think, have been strongly motivated to carry on where he left off. I know that I, for one, have always tried hard to keep my eyes turned in the general direction of those elusive stars he kept talking about, but it has never been an easy quest.

Gordon, on the other hand, was undaunted by the implications of a life dedicated in large part to the creation of an impressive mountain of dung.

"Nobody is too good to shovel shit," he declared with conviction.

"My colleagues at Dalhousie are getting worried about the amount of time I spend in the barn," I said. "They think I should be in the library doing research."

"Your colleagues are insane," he said simply. "They'd all be better off if they spent a little time working on a farm themselves. Or in a foundry, or a coal mine, or whatever. They ought to get out of that ivory tower of theirs once in a while. And anyway, why should your colleagues care what you do with your spare time?

"I guess they think I'll never get any publishing done if I spend too much time with the critters. It's publish or perish, you know. They're only trying to be nice."

"I wouldn't bet on it. But did you tell them you're averaging a chapter every two weeks on your textbook? What more can they expect?"

"I suppose they won't really believe I'm getting anything done till they see it in print."

"They'll see it," said Gordon with finality. "Meanwhile, just keep shoveling. It'll make you a better teacher, and a better scholar, too. It'll give you perspective. Besides, success isn't just a question of tenacity. It has a lot to do with being willing to tackle the dirty work. And that, I'm afraid, is what your professor friends don't seem to understand. But never mind. We'll just keep loading it on."

As I leaned my weight into my shovel that night I had visions of our half-frozen manure pile expanding relentlessly, one plop at a time, until it reached my father's ever beckoning stars.

The burgeoning barnyard

Gordon and I awoke one morning to discover that the grass had turned green overnight. After several false starts and numerous changes of mind, spring had finally arrived. The air was warm and redolent of freshly turned earth, and even the sparrows on the telephone wires seemed cheered by the new season. Gordon wasted no time in donning boots and overalls ("overhauls" as Travis called them), and he set off for the barn with a purposeful stride.

Valerie and I followed along behind him at the pace

of an average tortoise, for ever since she had learned to walk she had flatly refused to be carried to her various destinations, preferring instead to propel herself along on her own two chubby legs. Every few inches of the way her attention would be captured by a variety of inanimate and animate objects such as pebbles, twigs, leaves, beetles, worms, and the occasional slug. These she would examine with the intense concentration of a pawnbroker or a jewelry appraiser, but I had to make her let go of the less appetizing of her prizes or she almost certainly would have seized the opportunity to pop them into her mouth for a thoughtful taste test.

Gordon had already finished feeding the critters on the left side of the barn when Valerie and I finally made our appearance. I picked up a shovel and was about to scoop up the first offering of the day when I noticed that the barn floor seemed to be sprinkled here and there with what looked like grains of rice. I stooped down and peered at the rice more closely. Valerie observed me with great approval, thinking, no doubt, that I was at last catching on to the joys of scrutinizing and handling small objects to be found on the ground. To my horror I suddenly realized that the grains of rice were alive. Valerie was already delightedly squashing one between thumb and forefinger.

"Valerie, no!" I cried. "Drop it, Valerie!"

The maggots were everywhere. We scooped them up and hosed them out to the best of our abilities, but we knew they would soon be hatching into flies whether they were inside or outside the barn. Our first reaction was to buy some sort of pesticide to kill them off, but Travis had other ideas.

"No poison is goin' to kill them maggots any better than a flock of chickens. Now, I ain't the kind of man that tells his neighbors what to do. But if I was you lads I'd go out and get about a dozen hens. They'd keep the maggots clear of the barn, and they'd eat the bugs and weeds in your garden, too. Won't cost you hardly nothin', neither. They'd peck around and pick up the spilled grain that you shovel out anyway. And besides that you'd have nice, fresh eggs every day, at no extra charge."

Travis was so convincing that we went out that very day to visit a farmer who kept an assortment of barnyard creatures in a nearby community called Back Centre. It took us almost an hour to locate his farm, for we became hopelessly lost in the neighboring areas of Front Centre, Left Centre, and just plain Centre. Eventually we found him trudging down a muddy road, an arthritic old gentleman in a green cap and overalls, leading what looked like a slightly oversize deer with the horns and udder of a milk cow.

"You like Jerseys?" he asked as we fell in with him. "Them's the best kind of cows you can get, but nobody wants 'em no more. All the farmers around here have switched to those black-an'-white Holsteins you see yonder." He shook his gray head sadly and scratched his grizzled chin.

"Really?" I said, curious to know why anyone would prefer Holsteins to the beautiful, fawn-beige animal who now stood pointing her ears in our direction.

"Yes, the dairy commission pays by weight now, so of course everyone's doing all they can to get as much milk as possible from their cows. Them black-an'-white elephants give almost twice as much milk. But me, I wouldn't pay you two cents for it. Whitewash in a bucket, I calls it. You throw a penny in a bucket of Holstein milk and I guarantee you'll see the penny right through the milk on the bottom of the pail."

He spat in disgust at the very thought.

"Actually, we came to buy some chickens," said Gordon politely.

"Them Holstein farmers ain't savin' nothin', neither," muttered the old man as he opened a shed door and led the cow inside. "Them Holsteins may give twice the volume, but they eats twice as much, too. You can't get nothin' for nothin', and you're crazy in the head if you thinks you can."

Gordon agreed that getting nothing for nothing was an exercise that held very little appeal.

"This here cow's for sale, you know," he went on, closing her into a pen filled with fresh bedding and sweet-smelling hay. "She's the gentlest cow you'll ever see. Only four years old, an' due to freshen in July. I just dried her off this mornin'."

"So she'll be having a calf in July?" I said, looking at her sympathetically. My own baby was due the same month.

"Yup. I bred her to a Jersey bull, so if you wants that cow you'll get two for the price of one. You couldn't do better than that. And she'll give you nice, rich milk for the family. A little Jersey like that is just what you want for a family cow."

"I'm afraid I don't know the first thing about milking a cow," said Gordon.

"Ain't nothin' to know," said the old farmer reassuringly. "You just put the stool down and you grab two teats and pull."

By the time we were ready to go home we had bought and paid for seventeen nattering hens, a surly rooster, two ginger cats, and Daisy the Cow. They soon made themselves familiar with their new surroundings, and once again we were amazed at how they all knew their own roles without needing any help or instruction from us.

"And all because we put the critters inside for the winter," mused Gordon as he stood watching the denizens of the barn go about their customary business. "The cattle attracted the flies that created the maggots that feed the chickens, and the grain attracted the mice that will feed the cats that have already made their nests in the hay. It's like fate. There must be some sort of barnyard theory of inevitability that explains the acquisition of all these creatures."

We had achieved an ecological balance similar to that of the old-fashioned family farms that had existed for centuries in agricultural communities everywhere. The only discernible difference between those other farmers and ourselves was that they knew exactly what they were doing.

We review our multiplication tables

It was late May before the land was dry enough for us to let the critters out of the barn. As soon as Gordon opened the first stanchion and flung wide the barn doors, the critters immediately divested themselves of their former placidity and began to display what to us seemed like a distinctly unbovine sense of excitement. They lumbered out of the barn, knocking over pails and shovels, bumping into foundation beams, tripping over troughs, slipping on their excrement, and creating a state of general havoc among the cats and chickens. Once outside, they blinked momentarily in the sunlight, then kicked up their heels and tore off in random directions. Soon they were locking horns and backing one another down in an attempt to reestablish the pecking order they had so carefully worked out among themselves during similar exercises the previous fall.

The only other recreation they had enjoyed in that particular season had centered around the attentions of a local Hereford bull. Travis had kindly arranged for him

to "run" with our herd during a three-week period span-
ning the end of September and the beginning of October.
In return for his room and board and a nominal fee, the
bull was to make sure that all our cows and heifers were
"taken care of," as Travis put it. Gordon and I had expected our guest to be a massive creature of menacing
proportions who would arrive on the scene snorting and
pawing the ground, but he turned out to be a mild-mannered adolescent whose growth potential was yet to be
achieved.

"Are you sure we have the right bull for the job?"
Gordon asked Travis, looking dubiously at the awkward
teenager descending from the cattle truck. But before
Travis could answer, the animal had lifted his nostrils to
the wind and was already trotting with unswerving purpose in the direction of Prima, a large Black Angus who
had long ago browbeaten her sisters into accepting her
as head cow. The eager young bull circled her two or
three times with obvious relish, but he seemed unable
or unwilling to make his move.

"Maybe we should avert our eyes," I suggested.

"I think we should get a milking stool for him to
stand on," said Gordon.

"Now don't you be pokin' fun at him just because
he's not yet all growed up," scolded Travis. "The semen
from a young bull is the same as what a mean and cranky
old bull has to offer, but you won't see this one breakin'
down your fences or drivin' his head through the barn
wall."

The little bull was as good as Travis's word. By mid-
October he had become intimately acquainted with all
the "girls" without committing any acts of violence against
our persons or property, and by June of the following
year the results of his visit began to appear here and
there in the meadows. Gordon and I leaned contentedly

on the fences, watching the newborn calves discovering the world around them while their various mothers kept an anxious eye on them.

By early July our herd of critters had almost doubled, and our human family had increased by one third. But it wasn't until we brought our new daughter, Victoria, home from the hospital that we realized we had done everything wrong. Travis lost no time in apprising us of our latest missteps.

"You got eleven new calves standin' out there on the land, and ten of 'em is bull calves. And now the missus comes home from the hospital and she brings a baby girl with." He looked at Gordon disapprovingly. "Farmers is supposed to have *sons*, and if you wants to increase your herd what you needs is *heifer* calves. You city folks always gets everythin' turned around backwards. Ain't nothin' you can do with them bull calves 'cept beef 'em."

"You mean *slaughter* them?" I said, instinctively tightening my grip on Vicki, who was sleeping peacefully in my arms.

"They has to be slaughtered if they're goin' to be eaten, but you'll be waitin' two and a half years before they grows up big enough to take to market."

"Two and a half years," Gordon groaned, slapping his hand to his forehead. "Just think of the manure pile!"

"Are we going to have to kill them ourselves?" I asked apprehensively.

"No," said Travis, flashing his lopsided smile. "When the time comes, Mervin Zinck will take 'em away. He has all the equipment. The blocks an' tackle, and the tiled floors, and the coolers to hang the critters in . . ."

"Okay, okay," I said, hoping to stop him before he could paint too vivid a picture of the terrible future that awaited the innocent calves. All our animals had been multiplying at an incredibly energetic rate, and I hated to think that most of them would end up being devoured by hungry, carnivorous bipeds, some of whom would be members of the Jones family itself.

Travis, on the other hand, seemed unperturbed by the animals' sad destiny. His vision of the world differed sharply from ours, for his years on the farm had taught him to view life from an entirely practical standpoint. Nothing brought this home to us more clearly than the time when Gordon and Travis and I were standing on the shore after a long day's work in the fields. A heron came gliding in for a landing in the shallow water a few yards down the beach, and Gordon stood watching in admiration.

"Just look at that heron, Travis," he said, as it settled down knee-deep in the water. "Isn't he a marvelous bird?"

Travis looked at Gordon with an expression of immense pity for the poor city boy who could never seem to get anything straight.

"Jesus Christ, Gordon," he said, sucking quietly on his pipe. "There ain't enough meat on that thing to make a jeezly *hamburger!*"

4

Daisy the Cow

Daisy Jones versus the Dow Jones

We knew in advance that Daisy's calf, the last in the season, was bound to be a heifer. I looked out the kitchen window one morning and there she was, happily nursing away while Daisy gazed placidly at the morning sky. She was tiny and frail, with spindly legs and knobby knees, but it was easy to see she was a fine, healthy calf by the way she mercilessly butted her mother's udder in an effort to force the milk to flow faster. As a nursing mother myself, I couldn't help but wince.

"Well, look at that!" exclaimed Gordon, peering over my shoulder. "She had her calf during the night! And it sure looks like a heifer from here. I knew Daisy wouldn't let us down!"

Bambi was still nursing as we approached her to have a closer look. Her tail was flicking around at a furious rate while she eagerly guzzled down her mother's warm milk. Daisy turned her head and gave us a cursory glance before deciding to move on to a more promising part of the field. In so doing she pulled her teat unceremoniously from Bambi's mouth and left her standing there, legs still splayed, looking rather taken aback by the sudden disappearance of her lifeline.

"You lads is goin' to have to separate them two right away," said Travis, looking disapprovingly at Bambi as she gamboled around her mother. We had learned that Travis could always be counted upon to appear on the scene whenever something new was happening at our farm. His own farm was situated on a promontory that afforded him a comfortable view of our land, so little ever happened *chez Jones* that escaped his neighborly gaze. There were times when we viewed this as something of a mixed blessing, but on the whole we felt that the cordiality and helpfulness of rural people more than made up for the inevitable lack of privacy one experienced in the country.

"If you lets the calf suckle her mother twenty times a day," Travis went on, "she'll give her mastitis. Every time she sucks on a teat she leaves her open to infection. You should take her to the barn and wash off her udder real good."

We both had to agree that we certainly didn't want a cow with mastitis, so Gordon went and got Daisy's halter and put it over her head. But when he tried to lead her to the barn he soon discovered that she had no intention of following him. She took the time to give him one short glance of benevolent disdain, then she simply lowered her head to the ground and continued to graze as though her halter existed only in Gordon's imagination. She would allow him to lead her a few paces while she munched on the grass, but as soon as she was ready for another mouthful she stopped short and put her head down again. There was no way to move her by even so much as an inch until she was satisfied that she had enough grass in her mouth to last another four or five steps. I could hear Gordon muttering something about how it was going to take all day to get her to the barn at that rate.

"Now I know what Travis meant the other day when he told me that cows are nothing but trouble in a leather bag," he complained, glaring at Daisy.

"Here, let me give you a hand," I said, grabbing the end of the halter. "If we both tug at the same time, maybe we'll be able to convince her to come along."

The harder we pulled, the more Daisy resisted. We were both facing her and tugging with all our strength, but she remained firmly in place. She stretched her neck slightly in our direction, but this was the only concession she was willing to make to our combined sense of dignity. As I looked into her eyes I felt certain I could detect a gleam of triumph.

Then suddenly we both found ourselves flat on our backs on the grass. Daisy had unexpectedly stopped pulling against us and had managed to yank the halter out of our hands. By the time we scrambled to our feet she was halfway to the barn, her udder dancing wildly as she trotted along with amazing alacrity. It took us a few moments to realize that it was Travis who had come up with the solution. He had simply scooped Bambi into his arms and had headed directly for the barn. He had just opened the barn door when Daisy caught up with him.

"If you wants 'em to go somewheres or do somethin'," said Travis when we finally joined him inside, "you has to give 'em a *reason*."

"Employee motivation," said Gordon knowingly. "I should have thought of that myself."

"Cows is like anyone else," Travis went on. "They wants to have their own way, and your job is to get to know 'em so you can outfox 'em. Now take this here one. You don't want to leave her standin' like that in the middle of the barn because she'll make a mess right there on the floor. So we has to get her head through that there stanchion, and face her around so her rear end is over the trough. Any ideas?"

But before Gordon could reply Travis had filled a can with some of the delicious, sweet-smelling short-feed we had purchased at the Farmer's Co-op; then he went and dumped it into the manger of Daisy's appointed stall. Daisy, who up to that moment had been hovering solicitously near Bambi's pen, suddenly lost all her maternal instinct at the first sound of the falling short-feed. She shot into her stall and plunged her head through the stanchion just as Travis snapped it shut, and there she remained, quiet and uncomplaining, licking up the short-feed with her long, rough tongue.

"Well, you just gave Daisy an incentive bonus," said Gordon. "Otherwise known as a bribe where I come from."

"Cows ain't above takin' no bribe," said Travis soberly. "And they ain't above stealin' from their neighbors, neither. But right now I think you'd best put the milk stool down there and get to work milkin' out her swollen udder before that there short-feed gets gone."

At first Daisy was very patient about Gordon's trainee status, but when she finished her short-feed she decided she was fed up with having her teats pulled and squeezed by such a rank amateur, so she gave him a quick sideways kick that sent him tumbling off his stool.

"Are we going to have to bribe her again?" said Gordon indignantly. "That stuff's expensive!"

"If you was all alone I'd say you'd have to give her more till you gets your speed up some," said Travis. "But I'll finish her off this time. You just watch an' learn."

Travis settled himself confidently on the milking stool and leaned his head against Daisy's flank.

"That keeps her off balance so she don't kick," he explained.

Soon the milk was streaming rhythmically into the pail in long, even squirts, ringing musically as it hit the stainless steel sides. Daisy was immediately aware that she was in the hands of an expert. Not once did she raise a hoof or even swish her tail while Travis was milking her.

"A cow is really an amazing animal," commented Gordon as he watched Travis slowly fill the milk pail. "When you come right down to it, Daisy is sort of the rural equivalent of the New York Stock Exchange, you know what I mean? If you look at her in terms of stock certificates, her milk becomes dividends, and every year we get a stock split when she calves, and eventually our initial investment will be returned when we have to sell her for beef. What more could you ask from a cow?"

Daisy's only reaction to Gordon's assessment of her purpose in life was to give a long, drawn-out bellow, which was quickly answered by a plaintive little bawl from Bambi.

We learn about milk quota

The first pail of milk that Gordon brought to the house was a source of considerable excitement to both of us. We poured it carefully into twelve old-fashioned quart bottles that we had discovered in the Farmer's Co-op, then we stacked them all neatly in the kitchen refrigerator. Every once in a while we would open the refrigerator just to admire those glorious bottles of milk that had been collected, squirt by laborious squirt, from our very own Jersey cow. It was the first milk we had ever seen that hadn't come from a grocery store, and to us it was nothing short of a miracle. After a few hours the top fifth of each bottle was filled with luscious cream the color of buttercups, so Gordon and I spent the rest of the afternoon sampling it on fresh strawberries, in hot coffee, and whipped on top of homemade scones.

For the next three days Gordon dutifully brought home twelve quarts of milk from the barn every morning and every night, and I began to spend more and more time in the kitchen frantically baking custard pies, incubating yogurt, and concocting eggnogs, milk shakes, and every known variety of creamed soup. We had homemade butter on our toast, on our vegetables, in our gravy, and even in our oatmeal. Our potatoes were doused with sour cream, as was our borscht, our Stroganoff, and our garlic spinach. We consumed vast quantities of yogurt in combination with everything I could think of, including granola, fruit salad, and various dessert liqueurs. Our favorite dish turned out to be something rather simple: a goblet of yogurt covered with Jersey cream and pure Canadian maple syrup.

"I guess twenty-four quarts of milk is a bit more than one family can handle," observed Gordon one night.

"I thought you'd never notice," I said, smiling weakly as I tried to swallow my third helping of blancmange.

Our refrigerator was already bursting at the seams, and we clearly ran the risk of being drowned in milk if we didn't soon find an outlet for Daisy's overly generous production. Unfortunately there were no neighbors within a five-mile radius who didn't have cows of their own. We couldn't even give it away.

"And we can't hope to belong to the Farmer's Co-operative Dairy," said Gordon, wiping some whipped cream from his lips. "They'd never send their truck over here for the milk of just one cow. And even if they did, we'd have to get a bulk storage tank to keep it refrigerated for them, and with the expense and all the cleaning it just wouldn't be worth it."

"What about that little dairy in Bridgewater that

we pass all the time on our way to the co-op?" I said. "Maybe they're small enough to let us deliver our milk to them."

"I'll call them first thing in the morning."

It turned out that the La Have Dairy in Bridgewater was a creamery specializing in butter and ice cream, so they were not in a position to accept any milk. But they were delighted by the idea of taking our rich Jersey cream, and they assured Gordon they would buy whatever quantities we could provide.

"It works out perfectly," said Gordon, with obvious satisfaction. "We sell the cream to the dairy, and the farmer down the road takes the skimmed milk for his pigs. No more problems."

That afternoon we carefully skimmed the cream off the pail containing the morning's milk and proceeded at once to the La Have Dairy to make our two-quart contribution to the local ice-cream industry. We were greeted by a short, rotund man in a white lab coat whose duty it was to test all incoming cream for butterfat. I had put our cream in my best stew casserole, a handsome vessel adorned with beautifully hand-painted vegetables, but as I handed it over to the technician I couldn't help noticing the huge stacks of sensible, five-gallon buckets that were no doubt being circulated among the professional farmers in the area.

"It isn't very much," I said apologetically as he took the casserole from me and placed it near a rack of test tubes.

"Don't you worry," he said encouragingly, lifting the lid and sniffing the contents with the widely flared nostrils of one whose olfactory sense had been sharpened by years of experience. "We're glad for whatever cream we can get around here."

He must have been satisfied with his preliminary sniff, for he quickly rolled up his sleeves and set about testing our cream for its fat content. He stirred it thoroughly and then mixed a measured amount with a dangerous-looking acid, and before long he was pronouncing it to be of excellent quality.

"Jersey cream is always the best," he said, beaming with evident pleasure. "We always welcome cream shippers who have Jersey cows. God knows there are few enough of them as it is."

"Did you hear that?" I said to Gordon, nudging his elbow.

"Chalk one up for the Jerseys," he said, giving my arm a squeeze.

"No, I didn't mean that. Didn't you hear him? He called us *cream shippers!* Isn't it wonderful? We've become full-fledged, bona fide cream shippers, that's what we are. We have a real place in the community now. We ship cream from Lunenburg to Bridgewater, and we transport it in my stew casserole. Doesn't it make you feel *proud?*"

"Now, if you'll just give me your quota number," said the technician, smiling benevolently.

"What quota number?" said Gordon, giving him a blank look.

"Why, your quota number issued by the Nova Scotia Dairy Commission," he said, tapping his pencil.

"Nobody issued us any quota number," said Gordon. "Were they supposed to?"

"Did you apply?" asked the technician, looking up at Gordon over his glasses.

"Nobody ever told me anything about needing a quota number. I had no idea!"

It was a sad drive back to Lunenburg that evening. The gorgeous, yellow cream was still sloshing around in the casserole, having been ignominiously rejected by the La Have Dairy for lack of a quota number. The technician had been just as disappointed as we were and had gently explained to us that if he accepted the cream the dairy would lose its license.

"You mean there are spies from the dairy commission watching us at this very moment?" Gordon had exclaimed, peering behind the test tubes with an expression of wonderment.

But the technician had been steadfast in his adherence to the rules and regulations and could not be persuaded to add surreptitiously our little bit of cream to his two-thousand-gallon pasteurization vat. It was almost as if our quotaless cream were in danger of contaminating the entire tank.

"You'll have to go to the dairy commission in Truro," the technician had advised us. "You'll be asked to fill out an application form, but I think I ought to warn you that there's hardly any chance you'll get a quota this year. It's all been filled up till next March. But go and talk to them anyway. Who knows? Maybe you can work something out. It's not my place to say."

We felt thoroughly routed and confused by a bureaucracy whose function was as yet unclear to us. It seemed hard to believe that some anonymous men in an office in faraway Truro should seriously wish to prevent a humble housewife from trying to get rid of a casserole of cream. I felt certain that once I came face to face with these men and explained to them all about Daisy and Bambi and our predicament with the kitchen refrigerator, they would pick up the phone at once and destroy forever all the unnecessary stumbling blocks that lay between our farm in Lunenburg and the La Have Dairy in Bridgewater.

The Nova Scotia Dairy Commission meets Daisy the Cow

The Nova Scotia Dairy Commission was run by an individual named Tom Murray. He turned out to be a relaxed, avuncular sort of man with kindly blue eyes and a shock of snowy white hair. He invited me into his office and listened patiently to my entire story, smiling sympathetically and frequently nodding his head as I described the nettlesome problems that had finally led to my visit to the dairy commission that day.

"I can well understand your feelings of frustration when the man at the La Have Dairy told you he couldn't buy your cream," he said, when at last I came to the end of my litany of griefs. "But he was quite right to refuse you. He really had no choice at all."

He then proceeded to explain that the dairy farmers of Canada would soon have the country awash in milk were it not for the controls set by the Canadian Dairy Commission. These controls limited the production of each farmer to the amount set by his individual quota. The national quota was reassessed every year and divided up into provincial territories based on the average consumption of milk and milk products, but once the quota decisions were made for that year they could not be changed. This meant that Daisy's case could under no circumstances be considered until March of the following year.

"Couldn't you make an exception for Daisy?" I pleaded.

"There can't be any exceptions," said Mr. Murray firmly. "If I made exceptions my office would be inundated with petitions from farmers wanting quota for other cows they would like to buy; then the whole quota system would eventually fall apart."

"What would happen then?" I asked, trying to imagine a nation whose quota system had fallen apart.

"If we produced more milk than we could sell or consume," he said, leaning back in his chair, "the price of milk would fall so hard that many farmers would go out of business. Then we'd be short of milk and the prices would soar, but at that point it's not so easy to put the old farmers back in business again. Or even new farmers, for that matter. Farming is a very expensive business, as you know yourself. So if we took the controls away we'd end up with brief gluts of milk followed by long shortages, and wildly fluctuating prices. Everyone would suffer. The farmer, the consumer, the dairy industry, and the economy as a whole. So if you and Daisy can just bear with me till next March, I assure you I will personally issue you the quota you need to sell your cream to the La Have Dairy."

I had come prepared for a fight, for a knockdown, drag-out fight if necessary. I was ready to plead, to threaten, to cajole, to argue, to rationalize, maybe even to slip in a little false logic here and there, but I never expected to be so utterly disarmed by the obvious soundness of Mr. Murray's argument. He and his colleagues at the dairy commission had clearly devoted a good deal more thought and energy than I had to the question of what was best for the nation in terms of its milk production. I began to feel ashamed that Daisy and I had taken up so much of his time.

"No, please don't get up," he said, raising his hand with an authoritative gesture that seemed incompatible with his gentle personality. "We still haven't solved your problem yet. What about Daisy?"

"But you have bigger problems to worry about," I protested, surprised that he should be willing to concern himself with one local cow when the entire provincial herd was no doubt mooing constantly for his attention.

"I'm sure you didn't come here all the way from Lunenburg to talk about *my* problems," he said, smiling affably. "Daisy is the issue here, and there are several things we could do about her, but not all of them would make sense in your particular situation."

He leaned forward and put his elbows on the desk, then very carefully placed the tips of his fingers together two by two, starting with his little fingers. Once he was satisfied they were all well in place, he rested his chin and forehead on his thumbs and forefingers respectively, then contemplated me thoughtfully for a moment before continuing his explanation.

"You could start by either discarding Daisy's milk or drying her up," he began. "But of course neither of those alternatives makes any sense from a financial point of view."

He had closed his eyes in his effort to concentrate more fully on what I assumed must be the growing complexity of the various possible solutions to my dilemma.

"That being the case," he went on, "you could look at it in another way, if you like. You could cut your losses and run, which means selling Daisy right away to a farmer who has quota to spare. Or you could beef her."

"*Beef* her?" I cried, horrified by the suggestion.

He opened his eyes and looked at me from either side of his fingers, but I couldn't tell if he was smiling or wincing. "I can quite appreciate, however, that an alternative such as the one I just described would be unacceptable to you and your husband," he added hastily.

"You're right about that," I agreed, still too taken aback to allow myself to be quickly mollified.

"I would say that the only thing to do," he continued, "is to sell Daisy's milk legally, which essentially means one of two things. You could sell it all to a pig farmer, for example. But once again, it wouldn't be a very profitable way of disposing of it, I'm afraid."

"I'm beginning to wish I'd never bought poor Daisy,"

I said, feeling more and more discouraged by the seemingly endless difficulties involved in owning a cow.

"Now don't you worry," he said, folding his hands on the desk in front of him and beaming openly. "I still have one more suggestion, and I think you'll like it. As far as I'm concerned, your best bet would be to make some sort of a dairy product. You could sell it to a store, or to your friends, or to whatever outlet you can find. Something like butter, for example, or cottage cheese if you like. Whatever you think you can do best."

"And you don't need quota to sell butter or cheese?"

"That's right. Our business is to control the production of fluid milk, but we're not concerned about manufactured products. In other words, I can't allow you to sell fresh milk or cream to your local dairy, but you can sell as much butter as you like to any store that wants to buy it from you."

"You mean, I'm free to make all the dairy products I want, and nobody is going to care?"

"Oh, I wouldn't exactly say that. The Department of Agriculture will care about sanitation, and the Department of Consumer and Corporate Affairs will be interested in standards, but these are hurdles that can be cleared. You shouldn't have anything to worry about if you're just making a little bit in the kitchen for your friends."

Just then the receptionist opened the door and leaned in.

"Mrs. Jones's baby-sitter is in the waiting room, sir. She wants to know if Mrs. Jones could step outside and nurse the baby now. It seems it's past her feeding time."

I looked quickly at Mr. Murray, half expecting him to want to issue me some quota for Vicki's milk consumption, but he was smiling softly to himself, his mind evidently on other matters.

5

Daisy Goes into Business

The first experiments

Yvette was a peppy, garrulous, middle-aged Québecoise who had opened up a health-food store in Halifax and was doing a thriving business. Her only problem, she had once told me, was that she could not get any good yogurt. So when the dairy commissioner in Truro advised me to make a product with Daisy's milk, I went straight to Yvette's health-food store to ask her if she would like me to provide her with some fresh, homemade yogurt for her customers.

"It's guaranteed to be creamy, luscious, and hopelessly irresistible," I assured her.

Her laugh was wholehearted and generous. "You do not 'ave to convince me! I would be 'appy to carry such 'opeless yogurt in my 'ealth-food store. My customers, they will devour it with much appreciation and appetite."

I bought a package of yogurt starter from her, and feeling rather flush that day, I decided to invest twenty dollars in a brand-new quart-size yogurt maker. The new one together with the old one I had at home would allow me to make two whole quarts of yogurt at one time. Enough to feed an army.

* * *

The next afternoon I proudly deposited twelve sixteen-ounce white plastic containers of yogurt on Yvette's counter.

"Ah, but this is *magnifique*," she exclaimed, snatching up one of the containers and opening the lid. "Regard the cream, the beautiful yellow cream! It covers softly the yogurt below, like a smooth blanket." Yvette looked at the cream with genuine affection. "But it is not sufficient that I regard this creation. It is necessary that I savor it with the mouth and the tongue." Her expression had grown sensuous. "Voilà," she went on, her dark eyes narrowing with anticipation. "I 'ave 'ere the spoon."

She took a white plastic spoon from a drawer behind the counter and held it aloft with a dramatic gesture, as though she were showing a large audience a rabbit she had pulled from a top hat. I could almost hear the drumroll as she slowly lowered her spoon into the exact epicenter of the container. She curled her spoon expertly into the yogurt and lifted it to her lips. She hesitated a moment, as though preparing herself for the gustatory experience; then she opened her mouth and gently deposited the yogurt on her expectant tongue. As she closed her lips I could see her eyes roll back in their sockets, and instantly her brow was furrowed with what looked like a combination of pain and ecstasy.

"Mmmmm!" she declared, without opening her eyes. "This is without doubt the best yogurt that my tongue has had the honor to wrap around." Suddenly her eyes opened and she fixed me with a piercing stare. "Or is it wrap *about?* I always have trouble with the prepositions. On, under, over, around, about, they all drive me *crazy*," she confided in a low voice. "But your yogurt, he does not drive me crazy. He drives me *wild*."

By the time I got home that evening, Yvette was already on the phone demanding more yogurt.

"It sold like the crazy fire."

"Like wildfire, you mean."

"Wild, crazy, it is all the same. My customers were in ecstasy! They cannot understand where is the rest. They come back tomorrow. You must bring me three, four times as much. Is it so?"

Gordon and I went to work right away with our two little one-quart yogurt makers, but after dutifully grinding it out for five days and six nights in a row, it became abundantly clear to us that we would never be able to keep up with Yvette's insatiable demands unless we figured out a way to incubate the yogurt in larger quantities.

"We need something big, and something we can clean well," I said.

"What about that tropical fish tank up in the attic?" Gordon suggested. "It'll just sit there forever if we don't use it."

"It's big enough, that's for sure. I guess if we insulated it with blankets it might be all right."

"Let's try it then. That's the only way to find out if it's going to work."

Gordon cleaned the tropical fish tank while I filled four eight-quart pots with Daisy's milk and put them into the oven to pasteurize. I kept them there until they reached 185 degrees Fahrenheit, then we cooled the pots individually in a sinkful of cold water. Once the milk was down to about 125 degrees, we poured it into the tropical fish tank in the hall. We waited until it cooled to exactly 114 degrees; then we inoculated it with yogurt culture we had prepared from the previous day's production. I covered the tank with a tray and wrapped it all up snugly in a sleeping bag, while Gordon put on his boots to go down to the barn.

Three hours later we stood gazing happily at our first batch of tropical yogurt. It had turned out perfectly.

"I have only one question, though," said Gordon, looking perplexed. "How are we going to cool this stuff?"

My heart sank. We knew from experience that if we disturbed warm yogurt it would fall apart into a sloppy mixture of curds and whey, so it was important to cool our batches first before packaging them in the white containers.

"Should we try to cool it in the ocean?" I said feebly.

"Do you know how much that thing *weighs?*" Gordon shot back. I didn't need to be told.

"Maybe we could take it down to the cellar and put it in the freezer."

"That might just work," said Gordon, brightening slightly. "Let's give it a try."

We had just bought a small chest freezer that was still mercifully empty. It took twenty minutes' worth of moaning, protesting, and snapping at each other to get the fish tank downstairs and into the freezer, but we did it. And it worked. The warmth of the yogurt kept the freezer from dipping below 32 degrees Fahrenheit, so there was no danger of the outside edges of the yogurt getting icy. Within a few hours it was ready for packaging, and we were both overcome with hubris.

"A chest freezer with a tropical fish tank inside!" I crowed, rubbing my hands. "Do you think the other dairies will start sending industrial spies to our farm to find out how we do it?"

We had not yet learned that the other dairies, by some amazing coincidence, were already familiar with the basics of making yogurt, and their equipment was not very different from our own jerry-built invention. They had massive, jacketed, stainless steel vats that allowed for the circulation of both steam and ice water for heating and cooling the milk, and they controlled this process with either push buttons or computers. We had just rediscovered a very crude wheel, and in our excitement we mistook ourselves for geniuses.

We buy some pigs to eat our mistakes

If the tropical fish tank inside the freezer represented Stage One in our yogurt development, the other health-food stores pushed us into Stage Two almost immediately, for they soon realized they were losing all their customers to Yvette. It wasn't long before they were on the phone asking us to supply them with Daisy's yogurt too, and we hadn't the heart to say no. It was quite obvious, however, that we would not have room in our basement for any more fish tank–freezer combinations, so Gordon decided it was time to bring his management-consulting skills to bear on our expansion arrangements.

"It's not every day that a company triples its production in just two weeks," he said, beaming. "This is going to take *planning*."

Gordon's idea of planning was to bestow on me the lofty title of general manager in charge of research, technology, product development, and quality control, whereupon he turned me loose in the kitchen to discover new ways of making perfect yogurt in larger quantities.

"And what are *you* going to do?" I wanted to know.

"I'll be in charge of sanitation and raw materials."

Gordon easily dispatched his managerial duties by acquiring a heavy-duty shovel and a mild-mannered cow called Buttercup. My responsibilities, on the other hand, proved to be somewhat more complex and a good deal more frustrating than I had anticipated. I experimented with every conceivable way of incubating milk, tackling everything from the oven to the furnace room, but I was such a total failure that Gordon had to go out and buy six pigs to eat my mistakes.

They turned out to be enthusiastic and uncritical admirers of all my abortive efforts, greeting every bucketful of spoiled yogurt with shrill squeals of uninhibited delight. They quickly learned to recognize us as we neared their pen with the yogurt slops, for as soon as they heard us approaching they would scramble to their collective feet and stand right in the feed trough so that we had no choice but to pour yogurt all over their eager snouts and sturdy bodies. By the time the pigs finished squabbling and wallowing in the yogurt it had turned a muddy brown, but this did nothing to curb their voracious appetites. They kept right on grunting and snorting and shoving one another with piggish indelicacy until their trough was empty. Then they would look up at us with small, expectant, intelligent eyes until our immobile forms convinced them that no more yogurt was forthcoming, at which point they would plop down in the shade and fall asleep, twitching occasionally as they dreamed contentedly of their next yogurt feast.

"It's no good, Gordon," I said one morning as we were shopping at Canadian Tire for a new hose. "I've been working for three days straight trying to figure out how to deal with Buttercup's milk, but it's all ending up in the pigpen. I guess I shouldn't really be in charge of research and development."

"Don't be ridiculous. You're great with R&D. We just haven't hit on the right idea yet. But don't give up on it. What the hell, we *can't* give up now. We'll just have to keep on muddling along till we come up with the right solution."

"But it's so much work for nothing," I complained. "Haven't you noticed how haggard and disheveled I've been looking lately?"

He took a step backward and folded his arms. I sucked in my cheeks and tried to look wan. "Good God!" he exclaimed. "It's true! You're wasting away right in front of my eyes."

But his eyes were not looking at me. They were riveted on something behind my back. I turned to follow the direction of his gaze, but all I saw was some camping equipment next to a display of large styrofoam picnic boxes. I glanced back at Gordon, expecting him to explain his fascination with the camping gear, but he had buried his mouth and chin in one hand and was continuing to stare thoughtfully at the display.

"Forget it, Gordon," I said, feeling a bit exasperated. "We'll never have time to go camping. Anyway, summer's almost over."

"Those picnic boxes have been marked down fifty percent," he murmured from behind his hand.

"So what?" I said impatiently. "What's the use of even *thinking* about picnics? Besides, they're too big for us. There are only three of us, after all. Or three and a half, if you count Vicki."

"I was thinking of buying them all."

"*What?* Just because they're on *sale?*"

"Think, Sonia. Think. Can we do it? You're in charge of R&D."

"Incubators!" I cried, suddenly seeing the rather obvious connection. Two shoppers jumped slightly and glanced at me sideways. "Of course! We can use the styrofoam boxes as *incubators!*"

We bought eighteen styrofoam boxes for the price of nine and proceeded to cram them into our station wagon with the help of two bemused salesclerks.

"Boy," I could hear one say to the other as we drove away, "some people will buy just about *anything* for a bargain."

After several more days of experimentation we finally worked out a system of pouring the inoculated milk directly into small white plastic containers and incubating twenty-four at a time in a styrofoam box. We found that if we stuffed all the air spaces with packaging material the yogurt would set up perfectly in about three to four hours, provided the temperature was exactly 114 degrees when we filled the containers. This turned out to be quite a tricky procedure compared to our former method of batch incubation in the fish tank, for the milk tended to cool down too much as we were filling the containers. This yielded a gluey, overly mild yogurt with the conistency of a thick milk shake. We tried to compensate for it by inoculating the milk at a higher temperature, but the results were even more disastrous.

We finally learned to cool and inoculate only one stainless steel pot at a time, and then we proceeded to fill the small plastic containers as fast as possible, stuffing the airholes in the styrofoam box with anything at hand (tissue paper, dish towels, clean laundry, paper towels, newspaper). We then slapped on the lid and attached a piece of paper indicating the exact time of inoculation. We found that five boxes would take care of the milk production of two relatively fresh Jersey cows, and we quietly gloried in our newly discovered techniques.

There was cause for real celebration, however, when it occurred to us that we would not have to take out the containers after incubation. Instead, we learned to remove the assortment of airhole stuffings and replace them with frozen Fridgi-Paks, also available from Canadian Tire at a great reduction at that time of year. The frozen Fridgi-Paks would cool the yogurt overnight; then we would replace them with newly frozen packs the next morning and presto, the yogurt was ready to be loaded into our station wagon and delivered to the health-food stores.

"The people they are coming from all the corners of 'Alifax with joy and anticipation to receive your yogurt," said Yvette one morning. "It is with pride that I describe to them the 'ealthful qualities of the yogurt now available 'ere. Not all of my customers understand that the bacteria are manufacturing the vitamin B complex in their intestines. This is a marvelous system. The vitamin B complex is more easily absorbed if it is made this way than if it is swallowed inside a vitamin pill. I am talking my customers away from taking pills." She frowned and stood thinking for a moment. "Or is it that I am talking them *out* of it? These prepositions in English, they are most consternating."

Just then a portly, well-coiffed, gray-haired lady swept past us on her way to the bread counter. She glanced at us briefly through tinted spectacles and acknowledged us with an imperceptible smile. Yvette pounced.

"Ah, Madame Russell," she crooned. "You 'ave 'ere the fresh yogurt you 'ave requested. You cannot get it fresher. It was inside the udder of the cow of Madame Jones this very morning."

Mrs. Russell paused to take a better look at us over her glasses. "How very interesting. And did you milk the cow yourself, Mrs. Jones?"

"Oh no," said Yvette, before I could open my mouth to reply. "All the milking is done by 'er 'usband, Farmer Jones."

"Farmer Jones," mused Mrs. Russell. "How quaint."

"But it is Madame Jones who inoculates the yogurt with two separate strains of bacteria that are sent to 'er from a laboratory in my native Montreal," Yvette explained proudly. "And it is these same Montreal-born bacteria who live in your intestines and break down the fecal matter for you so that you never become constipated. Is that not marvelous? It is a wonderful gift they give to you, these little French bacteria, is it not so?"

"I never quite thought of it that way," said Mrs. Russell, trying to muster a polite smile.

"You see, as you get older the muscles of your intestines lose their effectiveness, and the peristaltic action must therefore slow down, causing the discomfort of irregularity."

Mrs. Russell was starting to sidle away, but Yvette was too engrossed in her explanation to notice. "This is why especially to my older customers I advise to eat yogurt," she went on, with her usual enthusiasm. "It is most beneficial, in more ways than two. Did you know that the bacteria in yogurt help to break down the cholesterol?"

Mrs. Russell was no longer there to answer. She had managed to make her getaway unnoticed while Yvette expounded on the benefits of yogurt.

"They break down the cholesterol," said Yvette, unfazed by Mrs. Russell's absence, "and they do it by the production of the vitamin B_6. This vitamin, as well as the choline and inositol and magnesium are most useful in this very important and 'ealthful process."

"Yvette," I said gently, "maybe you'd better not talk to people like Mrs. Russell about bacteria and constipation and things like that. I don't think she wants to hear about it. You might turn her off yogurt. Or you might even lose her as a customer!"

"Madame Russell?" said Yvette, laughing happily. "You must not worry, *ma petite*. Madame Russell 'as been coming to my store for many years, and every day I tell 'er something she does not wish to know."

Later that evening Gordon and I were walking down the dusty road to the barn, carrying buckets of yogurt that had perversely refused to set up properly in spite of our best efforts. The air was cool, and small clouds of gnats gathered around our heads without ever settling. The mosquitoes, however, were not quite so shy. They took advantage of the windless evening to take brazen nibbles from our necks and arms. Since we were each carrying two pails of yogurt slops we were utterly defenseless against their onslaught, and they seemed to know it.

"So Mrs. Russell thinks I'm quaint, does she?" said Gordon, trying to keep his mind off the dust, the sweat, the mosquitoes, the weight of the buckets, and the pungent smell of the looming pigpen. "That's one thing I've never been called before."

We tried our best to approach the pen as noiselessly as possible so as to get the yogurt into the trough before the squealing pigs could interpose their solid bodies and make a mess of the situation. But Argus, the pig who never slept, had his left eye open when we came into view. He leapt to his feet with a shriek of delight and made a dash for the feed trough while the other pigs, who evidently misunderstood the meaning of their companion's cries, ran around in total panic before zeroing in on the trough.

Gordon and I leaned over the side of the pen, swatting mosquitoes and watching the pigs feast on the failures of the day. Every once in a while they would shake their heads vigorously, flapping their ears and sending a shower of yogurt and flies into the air. The yogurt was the first to drop back into place, with the flies following close behind. As we stood there we realized that we were both beginning to appreciate the many different sides of farming—the hard work, the frustration, the dirt, the challenge, the feelings of accomplishment and self-sufficiency—but of one thing we were absolutely certain: life on the farm was anything but quaint.

A lesson in time management

By the time classes started again at Dalhousie in September, Gordon and I had established what we felt could safely be described as a reasonable semblance of order in our lives. Our two good-natured cows continued to provide us with all the milk we needed to satisfy the yogurt requirements of the various health-food stores, while I went on nursing Vicki night and day. Milk clearly had become the central issue for all of us.

In order to fit farming, teaching, mothering, and yogurt making into a twenty-four-hour period, we were forced to see to it that our daily schedule was a veritable masterpiece of time management. It became readily apparent to us that we would have to get up at four in the morning so that Gordon could milk Daisy and Buttercup and be back at the house by five. Meanwhile I would nurse Vicki, dress Valerie, and prepare breakfast. At five-thirty Gordon went out again to finish his chores while I cleaned up the kitchen and got the girls ready for the drive to Halifax. By seven o'clock Gordon had showered, shaved, and loaded the back of the station wagon with yogurt made on the previous day. We then turned on the radio and set off happily for the city, accompanied by news, weather, and Don Tremaine.

Gordon spent his days in Halifax comfortably ensconced with the children in Yvette's studio apartment, which happened to be located just two blocks from my office on campus. She was already at her health-food store by the time we arrived in the morning, and we were usually able to escape again before she came home at the end of the day. We made very sure we left everything exactly as we found it, and in return for her hospitality we would stock her refrigerator with fresh milk, homemade butter, and free-range eggs from our constantly roving chickens.

This arrangement turned out to be ideal for all of us. I would dash in to nurse Vicki and cuddle Valerie at various strategic moments of the day, and Gordon would have snacks and sustenance ready for us when I made my hasty appearances. All the pressures of the day would melt away as I shared news and Cabernet Sauvignon with Gordon on the third-storey veranda. The upper branches of an ample oak tree cast their shade over the glass table and wrought-iron chairs, and a cool autumn breeze would gently rustle the leaves as we sat looking out over the quiet, tree-lined street below.

"Well," said Gordon one morning as we sat watching the children playing on the carpet. "Is your supervisor happy now that you've finished writing *Spanish One?*"

"Not really," I said, thinking back ruefully on the events of the previous day. I had bounded gleefully into his office brandishing the completed manuscript of the textbook and placed it triumphantly on his disorderly desk. I had then stood back and waited patiently for him to look at it, happily imagining that I would soon be basking in the warmth of his unrestrained approbation.

"What's this?" he had growled, looking at the thick binder with an expression of growing alarm as it slowly occurred to him that it might represent some work for him to do.

"It's the book. The beginning Spanish textbook," I exclaimed, practically bursting with all but uncontrollable pride. "I told you I'd have it finished in a year!"

"You wrote a textbook in a *year?*" He looked up at me over his glasses. "But, my dear woman, nothing of any value can possibly be written in just a year."

My heart collapsed. "But it *is* good," I heard myself say in a rather defensive tone. "It's different from all the other Spanish textbooks on the market because I made it fun for the students. I created characters that remain constant throughout the book so there is a story line to follow. I class-tested it as I went along, and it works! My students had a great time discussing the material, and our classes really came *alive*."

"Let me tell you something, Mrs. Jones," he said, gently pushing the binder back in my general direction. "I once had a good friend who wrote a very excellent textbook. And do you know how long it took him to write each exercise? It took him anywhere from two to three weeks, and sometimes even longer. These things have to be given the proper thought, you know. They require a sustained and concentrated effort. They can't just spring to life all by themselves. Not if they are going to achieve any academic recognition. It just doesn't work that way, Mrs. Jones."

It infuriated me that my supervisor took it for granted that the manuscript of *Spanish One* was worth only the most cursory of all possible glances. I had taken infinite pains to see to it that no grammar concept or vocabulary word crept into any of the sections without first being explained in the appropriate chapter, and I had also gone to great lengths to introduce a large number of carefully selected cognates so that the students could expand their vocabulary quickly enough to be able to discuss fairly sophisticated themes right from the beginning of the book. Creating the dialogues and prose passages had been tantamount to putting together a collection of giant crossword puzzles, but the results had been well worth the effort. For the first time in my eight years of teaching experience I felt I had a textbook that allowed the students to progress rapidly and accurately, without their being subjected to reading material that was either confusing or boring.

"How many typed pages do you have there, anyway?" muttered the supervisor, unexpectedly taking back the binder. "Why, it's over eight hundred! How on earth do you think your students are going to be able to pay for photocopies? It'll come to more than eighty dollars. And besides, you're going to have to write a workbook to go with the text, and a program for the language laboratory. Your students will never be able to afford a package like that. It could cost them a hundred and fifty dollars or more!"

"I'm not going to hand out photocopies," I smiled. "Van Nostrand is going to publish it. Textbook, lab manual, workbook, and all."

"You've found a publisher?" he said, unable to hide his astonishment. "That's incredible! I would have thought the market was flooded with Spanish textbooks."

He was right. In the months before we moved to Nova Scotia I had sent out sample chapters and a statement of purpose to fifty-two publishers and had been rejected by every one of them. I had started with the best-known houses for the first few weeks; then finally I had sent multiple submissions to the rest of the publishing companies in alphabetical order until I had ended up with Van Nostrand. The senior editor there was too busy to explain in his letter the reasons for rejecting my manuscript, so he extended a halfhearted invitation to stop in and see him if I was ever in the neighborhood. I made it my business to visit him the very next day, and he irritably agreed to give me five minutes of his time. We ended up talking for two hours about how to slant a textbook for today's market and how best to package it for maximum appeal. The rest had been up to me.

"I wish you the best of luck with it," said the supervisor, smiling thinly. "There are any number of textbooks on the market today that enjoy considerable, how shall I say . . ." He hesitated as though trying to decide how to dispose of a very small dead mouse. "That enjoy considerable commercial success. But in my experience I've found that the real academics, the *true* scholars, are very rarely on the receiving end of anything substantial in the way of royalties. And as for churning out a textbook in little more than a year, well, I've already voiced my opinion on *that* subject."

Gordon chuckled when I told him the story. "Poor old fellow," he said sympathetically. "Look at it from his point of view. He's been a professor for most of his life, and what has he got to show for it? He was given tenure back in the days before anyone was expected to publish anything, and now he's a fixture at Dalhousie. Then along comes a young hotshot like you, and the first thing you do is publish a book. No wonder you rub him the wrong way. It probably takes him three weeks to tie his shoes. Anyway, we'd better get a move on now, or Daisy and Buttercup will be giving me dirty looks when I show up for milking this evening."

*   *   *

We would arrive back on the farm at five-thirty or six o'clock, whereupon I would cook dinner while milk simmered (and occasionally boiled over) on the stove. Valerie and Vicki were generally in bed by eight o'clock, which gave me just enough time to correct papers or prepare classes for the next day. As for the textbook, it did not, as my supervisor feared, spring to life fully armed like Athena from the forehead of Zeus. It was written in far more mundane circumstances, amid pots and pans and pigs and more cases of Pampers than I care to enumerate. For I had learned an invaluable lesson from a German roommate I had once had as a student in Paris. She was the very essence of Teutonic industriousness and she taught me to have an enormous respect for the five- or ten-minute interval. Before I met Ulli, I was of the firm opinion that it was altogether impossible to get down to anything unless I had at my disposal countless hours of uninterrupted time, but Ulli taught me otherwise. She would study in the Metro, between classes, in the cafeteria, on buses, and even in the dentist's chair, and no amount of noise could break her unshakable concentration. Although I have never come close to reaching Ulli's pinnacles of supreme assiduity, I did learn to come to terms with my old habit of procrastinating.

And so it was that Gordon and I kept muddling along on the farm, juggling cows, chickens, yogurt, styrofoam boxes, babies, and scholarly articles from one end of the day to the other, like working fools. We never had time to ask ourselves if we could have been missing some of the pleasures that come with a more leisurely existence, nor did we ever stop to remind ourselves that one of the reasons we had left New York was to escape from the hectic pace of the city.

6

Getting off the Ground

Peninsula Farm is born

By late fall our fields were almost bare except for an occasional tuft of yellowing timothy. The critters had come to the end of their summer gnawing program and were standing around mournfully looking for something more substantial to eat. They would huddle patiently together with their backs to the chill wind, while high above them the last geese of the season flew in perfect formation to winter homes in warmer climes.

It was on just such a day that a blue Mercedes-Benz pulled slowly up our muddy road and stopped irresolutely near the front door. Gordon was outside at the time, pitching hay to the hungry critters in hopes of postponing for as long as possible the unwelcome day when he would have to round them up and put them into the barn for the winter. A well-dressed man emerged from the Mercedes and began picking his way gingerly around the ruts and puddles as Gordon strode quickly toward him. Soon the two men were wandering over to the house, enjoying an animated conversation and stopping at frequent intervals along the way to laugh and gesticulate as they exchanged information. Our visitor was still chuckling to himself as Gordon ushered him into the kitchen.

"Well, I think we have a yogurt customer here," said Gordon, helping him off with his coat.

"You're welcome to all you want," I said, turning to the refrigerator. "I think we have at least seven or eight containers left."

"I don't believe he's talking about just a few containers," laughed Gordon. "I think he wants a truckload."

"A truckload!" I exclaimed, stopping dead in my tracks. "What kind of a truckload? I mean, how big a truckload? And when?"

"Now don't get excited," said Gordon mildly. "This gentleman is Mr. David Sobey, and he's the president of the Sobeys supermarket chain here in the Maritimes."

"How do you do," he said, extending his hand.

"I'm not sure," I smiled, returning his firm handshake. "If you want me to send yogurt to all your stores, there's no guarantee I'll still be alive in the morning."

He threw back his head and laughed good-naturedly. "No, no. Nothing like that. There are some stores in Halifax, though, that would be very happy to carry your excellent yogurt, but only in the quantities you think you can comfortably handle. I can see you have certain manufacturing limitations in your present facilities." He looked discreetly around the kitchen.

"I think Mr. Sobey expected to find at least a small dairy plant here, with maybe a smokestack or two," Gordon put in.

"I did indeed," agreed Sobey, "although I must say I was quite surprised that I had never heard of your company before. I happened to come across your product in one of the health-food stores, and a very charming woman named Yvette told me in no uncertain terms that yours was the best yogurt anywhere. Once I tasted it I realized the woman was absolutely right, and that's when I decided I had to carry it in my chain. I'm always very interested in supporting local businesses, but this was one I had never come across before."

He beamed at me with kindly, thoughtful eyes, little suspecting that it had never occurred to me to think of myself as any kind of a business at all. He seemed to take it for granted that we would be expanding, but when *I* thought of expansion I had visions of yogurt oozing uncontrollably from every nook and cranny of the kitchen, working its way throughout the house, absorbing everything in its way like a white alien blob from a science fiction movie.

"What is the name of your company, by the way?" Sobey inquired.

"Our company?" I said, looking blank. Sobey nodded and smiled pleasantly. I glanced over at Gordon to see if he had any ideas, but he was obviously in just as much of a quandary as I was. The very thought of our being an actual company struck me as rather comical under the circumstances, since our entire inventory of equipment consisted of ten stainless steel pots, eighteen styrofoam boxes, thirty-six Fridgi-Paks, and a very reliable kitchen stove. Sobey could see how we were operating and what our minuscule production capacity was in our kitchen, and yet he was standing there before us, amused and respectful, making us a perfectly reasonable business proposition. He had faith in us, and we couldn't let him down. We had to have a company name.

"We call ourselves *Peninsula Farm*," I said, trying to cover my confusion with an apologetic smile. It was the first name that came into my head, since our farm was located on what is known as First Peninsula. I hoped Sobey would not think our company name was absurdly unimaginative.

He absorbed the information with equanimity. "Peninsula Farm. Very good, very good," he nodded approvingly. "Now, when you deliver your yogurt to our stores, you'll have to label it with your company name, of course, and a list of ingredients. You'll need to do this to satisfy the requirements of the Department of Consumer and Corporate Affairs, which looks after labeling in this province. Those plain white plastic containers of yours are all right for the health-food stores, but when you get into the chains they're going to want to see the proper labeling."

"Don't worry," said Gordon heartily. "We'll have them labeled and delivered with no problem at all. You can count on Peninsula Farm."

Sobey smiled. "Then it's all settled?"

"You've got yourself a deal," said Gordon, shaking Sobey's hand. "Right, Sonia?" he added, raising his eyebrows in my direction.

"Right," I murmured. After all, it wasn't every day that the president of a supermarket chain came to our kitchen asking for our yogurt. How could anyone resist such a gesture? And anyway, he only wanted the product in a few stores to start with. A few more stores couldn't possibly make that much difference. . . .

"You know," said Gordon thoughtfully after David Sobey left, "this whole thing could turn out to be something really interesting. He made us an offer we couldn't refuse. You realize that, don't you, Sonia? To grow at the right pace, to be welcomed into the stores at our own timing—it's the sort of thing any businessman dreams of. It's a wonderful opportunity!"

"But how are we ever going to squeeze it into our schedule?"

"We'll manage somehow. It may take some vigorous muddling, but we might be able to figure out a way to keep farmers on the land by showing them how to combine farming with business or manufacturing. You said yourself you were concerned about land being taken out of production and sold for real estate. Well, if we could help farmers make a decent living maybe we could reverse the trend. Who knows? It could be a really exciting experiment!"

"I can see it all now," I smiled. " 'Brooklyn boy makes good on Nova Scotia farm.' Or 'Brooklyn yogurt maven revolutionizes Nova Scotia farming.' You tell 'em, Gordon."

"Hold it, now. It's not just up to me to make this company fly. When we incorporate, I think you ought to be the president. I'm more or less tied down on the farm, but you're in and out of Halifax all the time, so you're in a better position to be the spokesman and talk to the executives. Besides, I was the president of my company in New York, so now it's time for you to give it a try."

His eyes lit up with entrepreneurial glee as he continued to describe the various directions our budding enterprise might take and all the potential for good that lay inherent in the idea that was slowly taking shape. As Peninsula Farm began to take root in his imagination, our destiny seemed clearly to be veering away from Gordon's dream of sailing the seven seas.

The official taste test

"The supermarkets are a whole new ball game," said Gordon as he sipped his morning coffee. "The kind of people who shop in a supermarket are nothing like your average health-food customers. They're going to want flavors. You know, something like raspberry, or peach, or strawberry." He put down his mug and gave me an appraising look. "You're in charge of product development. Why don't you roll up your sleeves and get going on a couple of recipes? I bet it wouldn't take you five minutes to come up with a strawberry yogurt that would knock their socks off."

I looked at him warily. "I was going to spend the morning typing a scholarly article that I'm submitting to a journal."

"I'll do that for you," he volunteered. "You shouldn't be wasting your time typing. You should be doing important things, like developing yogurt recipes. I'll have the typing done by lunchtime if you promise to give me some strawberry yogurt for dessert. Okay?"

"I have to nurse Vicki first."

"No problem. I tell you what. I'll go to the store and get you all the ingredients you need, plus I'll buy you a container of every one of the strawberry yogurts put out by the other brands. Then we'll have a taste test and we'll see which one is the best."

Soon the kitchen table was covered with items to be used in the forthcoming experiment in the development of a brand-new strawberry yogurt for the people of Nova Scotia. There were several containers of Peninsula Farm yogurt, a bowl of defrosted strawberries, a canister of sugar, a handful of plastic tasting spoons, one large mixing bowl, a stack of empty yogurt containers, and a glass measuring cup. I washed my hands and donned an apron, half expecting to hear a drumroll or perhaps some dramatic background music, but the only sound to be heard was the monotonous clack of Gordon's dogged typing coming from the spare bedroom upstairs.

The rest was simplicity itself. Gordon was right when he calculated that it would take only five minutes to develop the recipe that was to remain essentially unchanged throughout the years. I had only to dump some Peninsula Farm whole-milk yogurt into the mixing bowl, stir in the freshly thawed strawberries until the color looked appealing, and add sugar to taste. The results were enough to make even the most modest manager of dairy-product development brag just a tiny bit, although it was obvious to me that 99 percent of the credit belonged to the ambrosial quality of the fresh milk and uncooked berries. Yet in spite of my not wanting to usurp the glory that properly belonged to the ingredients themselves, I couldn't help bursting out with a loud cry of "Eureka!"

There was a pause in the typing upstairs.

"Did you sneeze down there?" came Gordon's voice finally.

"Come to the kitchen and try our new Peninsula Farm strawberry yogurt!" I called.

"I'm only on the seventh page," came Gordon's reply. "I still have nine more to go. Don't you want me to send your article off to the journal as soon as possible?"

"Never mind the journal. You were right. Yogurt is much more important. Wait till you taste it!"

Since Valerie was the only member of the family who had any real objectivity about the quality of our competitors' yogurt, we decided to put her in the highchair and make her the official taste-tester. Her reactions were flawless. Every time we put a spoonful of a commercial brand into her mouth, she flared her nostrils and delicately pushed the yogurt out with her tongue.

"God, this is sickly sweet," said Gordon, wrinkling his nose at one of the samples of the commercial yogurt.

I looked at the ingredients. "No wonder. There are three different types of sugar listed. Glucose, fructose, and sucrose. They list them all separately so they can mention them *after* the fruit. That way the customer thinks there's more fruit than sugar, when it's really the other way around."

"Yes, well what gets me is that this brand is touted as a low-calorie yogurt because it's made from skimmed milk. But if you look at the list of calories you can see they're almost exactly the same as the other brands. The difference is hardly worth mentioning."

Gordon tried to fit a spoonful of another brand of strawberry yogurt into Valerie's mouth, but she wasn't having any. She kept turning her face from one side to the other, so that his attempts all landed on her pink, round cheeks.

"I think she's trying to tell us something," I ventured. "You'd better not give her any more of the commercial brands or she'll be turned off yogurt for the rest of her life."

"The kid has good sense," said Gordon, tasting what was left on the spoon. "This one has so little fruit in it I wouldn't know *what* flavor it was supposed to be if I hadn't seen the strawberries pictured on the package."

I stirred the yogurt until I finally fished out what appeared to be a piece of bleached strawberry.

"Here's one," I said triumphantly, holding it up for his inspection.

Gordon took one look at it and sneered. "Okay," he said, swishing some water around in his mouth like a professional wine taster. "Now it's time for ours."

I held my breath as I watched him put a spoonful of Peninsula Farm strawberry yogurt into his mouth. This was the first moment in the history of our company that anyone had tasted my strawberry yogurt. Drumrolls.

"Well," he said cautiously. "It's youthful, and a little frisky. Maybe even a trifle audacious, but it has character."

"Gordon, will you come off it? What do you *really* think?"

"You know what I really think," he said, grinning.

"Not one of the other brands even comes *close* to the recipe you put together. Yours is nothing less than phenomenal."

"You're coming dangerously close to tooting my horn," I said, feeling secretly gratified.

"Well, true is true. Why try to deny it? There's no use pretending to be modest about it. We have the best damn yogurt on the market, and that's a *fact*."

"You better not let anyone hear you say that. It sounds sort of vain, doesn't it?"

"Vain, schmain. You're in the yogurt business now, baby. You're going to have to get used to that. And if you're going to make yogurt, you might as well make the best yogurt in the world, or else don't even bother. The excellent, top-flight, successful companies are all proud as hell of their products, and they don't mind saying so, either."

"I guess I'm just suffering from a lifetime of conditioning to be modest. You know, the womanly woman, and all that stuff."

"Well, you can go on being as womanly as you like, but don't pretend you don't know you're the best. Being the best is the only thing we'll have going for us when it comes time to compete with the big boys. That, and being local. Have you noticed where all the rest of this yogurt is coming from? Just take a look at these containers. Yoplait is a French company operating out of Quebec, see what it says? And Light 'n' Lively is made by Sealtest, an American company with a branch in Quebec again. Crescent is made by another American company called Beatrice Foods, and they also have a branch in Quebec. Then there's Delisle, which makes Silhouette as well, and their headquarters are in Quebec too. So you can see what we're up against. The only dairy producing yogurt in Nova Scotia is Farmer's, and they can use all the help they can get. We shouldn't be shipping in so much yogurt from Quebec. It makes no sense at all. We should be making our dairy products right here, with local milk."

"That's right," I said, feeling more and more excited by the idea of starting a new Nova Scotian company. "And if we're successful, we'll provide some local employment, too."

"Of course. And not only for people, but for cows as well. Our motto should be 'Buy Peninsula Farm yogurt, and give a local cow a job.' "

"It all sounds great," I sighed. "But the truth is, we're just dreaming. I'm only a housewife with two cows and a kitchen stove, and we're talking as though we can go up against Sealtest and Beatrice Foods! Who are we kidding? They'll just laugh us off the map. They'll squash us like bugs if they ever feel like it."

Valerie banged her highchair impatiently, wondering when she was going to be getting some *real* food.

"Try giving her some Peninsula Farm yogurt," said Gordon. "She'll tell us better than anyone else whether or not we're going to be squashed."

Valerie looked suspiciously at the spoonful of our yogurt that I held out to her, but hunger got the better of her and she trustingly accepted my offering. This time she swallowed contentedly and opened her mouth for more. Before long the whole container of Peninsula Farm yogurt was gone.

Daisy is dubbed "Cow of the Year"

"Ah yes, the people from Lunenburg," said Tom Crowell, the manager of the Queen Street Sobeys store in Halifax. We had arrived in our station wagon to make our first delivery to the store that David Sobey had recommended

as having the most discerning and sympathetic customers. "Mr. Sobey told me you'd be coming in," Mr.
Crowell added. "Just haul your truck around to the back
loading area and pull up to the third bay. You can't miss
it. You'll see the empty milk boxes stacked by the door.
Go ahead and put your product on a pallet and I'll get
the receiver to let you in. I've cleared off a nice eight-
foot section for you in the dairy case. Mr. Sobey said he
wanted you on the eye-level shelf."

"Wait a minute," I called, as he strode away in search
of the receiver. He turned and looked at me question-
ingly. "We don't have a truck. We don't need a pallet.
And we couldn't possibly fill an eight-foot shelf!"

Mr. Crowell looked baffled. "Oh, I'm sorry. I thought
you were making a delivery."

I was feeling more embarrassed by the minute. "We
only brought two dozen containers along," I said, trying
to muster an apologetic smile.

"Oh, you just have *samples* today," said Mr. Crowell, beginning to see the light.

"Well, not really," I faltered. "It represents the excess yogurt of three days' production."

"Two cases of yogurt? That's all you have?" he said
incredulously. I could see a smile starting to form at the
corners of his lips, but Mr. Crowell, an obviously kind-
hearted individual, diplomatically suppressed it. "I'm afraid
two cases of yogurt won't be quite enough to satisfy the
customers in this store. Two cases of yogurt! I doubt
very much they'd last much longer than half an hour or
so. Come with me, I'll show you what I mean."

Mr. Crowell's stride was so long and his pace so rapid
that we were forced to break into a lively trot to keep
up with him. Several customers glanced up from their
shopping to watch our mad dash down the aisle in pursuit
of our speedy guide, but Gordon managed to dispel all

symptoms of alarm with a reassuring smile that he flashed indiscriminately in as many directions as possible.

"You see," said Mr. Crowell, who had been waiting patiently by the dairy case for us to catch up with him. "We've devoted about forty feet of shelf space to yogurt, and even *that's* not enough."

We stared respectfully at the impeccable display of yogurt that had been mass-produced by our competition in the provinces of Ontario and Quebec. Only one other brand had been produced in Nova Scotia, but it was similar in flavor and consistency to all the others.

"We sell more yogurt in this store than in any other outlet," Mr. Crowell told us, "so you people are going to have to hustle to keep up with the demand. If your product catches on, I'll need to keep the shelves well stocked with it, or I'll have customers complaining to me that they can't find what they're looking for. And you know what *that's* like."

Gordon stroked his chin thoughtfully as he watched the dairy case progress from a state of mild upheaval to one of undeniable decimation. "This is a situation," he remarked, "that definitely calls for another cow."

"Another cow?" echoed Mr. Crowell, his eyes crinkled with amusement. "You're going to buy another cow just for my store?"

"Well, maybe two cows," said Gordon, reappraising the dairy case with what he hoped Mr. Crowell would interpret as a highly professional eye.

"Now I've heard it all," said Mr. Crowell, throwing back his head and laughing heartily. We politely joined him with a subdued chuckle or two. "Now don't get me wrong," he added soberly. "I think it's terrific. I think the whole concept has enormous potential. I'll get a lot of mileage out of this. My customers will love it. They're tired of mass production, and they're sick of huge, anonymous conglomerates putting out plastic food. Whatever happened to quality, anyway? Listen, do you have a picture of your cow?"

"Who, Daisy?" said Gordon.

"That's her name, Daisy?" said Mr. Crowell. "I love it. Daisy the Cow. Look, all you have to do is get me a picture of her, and I'll hang it up here on the dairy case with her name under it. Or maybe I'll put it up on the wall where I keep the color portraits of my employees of the month. Daisy can be the "Cow of the Year." She'll be our mascot. We'll have the freshest yogurt in Halifax, and everyone will know it. Those outfits up in Ontario and Quebec can't possibly truck yogurt down here and have it come in as fresh as yours. Believe me, Daisy will attract a lot of attention. When it comes to dairy products, freshness is everything."

"I can't wait for you to try some of ours," I said, warming to his enthusiasm. "Shall we bring it in? It's just out in the car."

"That's right, you have some with you, don't you? Well, bring it in, then. You might just as well carry it in through the front door. It's not worth bothering the receiver to let you in the back."

When we returned with our two dozen yogurts we found Mr. Crowell busily stocking some shelves with jars of instant coffee.

"No rest for the wicked," he smiled as we drew near. "Joey's on break, and Bob's unloading a semi. But the coffee *must* go up. . . ."

"Take your time," said Gordon, impressed by Mr. Crowell's dexterity in handling the coffee jars. He was expertly snatching them from their cases and slapping them down on the vibrating shelves all in one blurry sweep of the arm.

"Now then," he said, turning to us when he had

finished. "Let's take a look at Daisy's yogurt." He plucked a container of strawberry yogurt from the box I was holding and spun it deftly in the air, somehow making it drop into his hand with the label facing forward.

I was proud of that label. I had made a special trip to the post office to buy labels with red airmail borders so that people could easily distinguish our strawberry yogurt from the plain yogurt (whose labels had no border at all). I had painstakingly typed on each one the information required by the Department of Consumer and Corporate Affairs, including our company name, the list of ingredients, the net weight, and a bilingual admonishment to the buyer to keep the product refrigerated at all times. I had even translated "Daisy the Cow" to "La Vache Marguerite" on the French side of the container. Gordon had supplied the finishing touch by stamping the "best before" date on the top left-hand corners. When our job was done, we had observed our handiwork with great pleasure, gloating openly at the thought of the fear and trembling that would strike the hearts of our competition when Peninsula Farm yogurt made its debut in the stores.

That was before we saw our containers actually lined up on the dairy-case shelf next to the commercial yogurt from Quebec and Ontario. Our sorry little white plastic tubs faded completely into the background, and the typed labels disappeared from view at the modest distance of five or six feet. At first I blamed my hopeless myopia—the outside world has a way of sinking irremediably into a shapeless haze whenever I take more than three giant steps backward—but on closer scrutiny I discovered to my horror that my precious airmail labels were not designed for clinging to plastic, especially at chilly temperatures. Many of them had already given up the ghost and were fluttering silently to the ground, only to be callously trampled by the hoard of eager shoppers jostling one another as they reached automatically for the brand-name yogurts they were accustomed to buying.

When I turned around I saw Mr. Crowell just finishing off one of our tubs of strawberry yogurt. He licked his lips with satisfaction.

"I'll say one thing. That yogurt of yours is the best product I've had in a long time. It's just like strawberry shortcake. There's no doubt in my mind at all. Peninsula Farm yogurt is going to take over the market. But let me give you a little word of advice. Get the self-adhesive labels. They're much better than the ones you have to lick."

"Of course," said Gordon, looking a little sheepish. "I don't know why we didn't think of that before. It was the airmail border, I guess. It seemed like the best way to identify the strawberries."

Mr. Crowell smiled. "Just keep on trucking. Someday you'll be able to afford a printed container like everyone else. Then there'll be no holding you back."

As we wended our way homeward through the afternoon traffic, I began to realize that my customary wariness was clearly giving way to a feeling of general, overall exhilaration. David Sobey believed in us, and so did Mr. Crowell. I was convinced that once our nondescript white containers were adorned with self-adhesive labels all the customers in the Queen Street store would be sure to buy one and take it home to their families, at which point children, parents, friends, and relatives alike would promptly faint in ecstasy as soon as our yogurt found its way into their mouths. There would be a mob at the Queen Street store. Mr. Crowell would have to build a protective barrier around the dairy case, and numbers would have to be issued to the eager customers. Peninsula Farm yogurt was destined to take Halifax by storm. We might have to buy as many as eight or ten cows, I thought, before we reached the saturation point. My feeling of exhilaration became tinged momentarily with anxiety until it occurred to me that there was room in the barn for a grand total of *thirty* cows, if there was enough demand. I settled back comfortably in my seat, secure in the belief that the milk of thirty cows was bound to last forever.

7

The Chain Reaction

Strong links

The necessity of filling the eight-foot dairy-case shelf that
Tom Crowell had so faithfully reserved for us in the
Queen Street Sobeys store led to the immediate purchase
of two more Jersey cows from a farmer in La Have. We
named one Clover, and the other we decided to call Melody in honor of Gordon's inexplicable urge, when he first
saw her, to play a melody on her prominent ribs. Although poor Melody was by far the skinniest cow in our
odd assortment of critters, she turned out to be our best
milk producer.

"Them bony cows always gives the most milk," said
Travis, looking approvingly at her generous udder. "The
feed goes to milk instead of fat. But you has to watch
out for milk fever when she freshens. Them cows with
the big udders goes down hard when they has a calf."

His words were prophetic. Two days after she had
her calf we found her stretched out in her stall, unable
to move, and looking for all the world as though she were
breathing her last. We called the vet in a panic. He soon
arrived on the scene and carefully injected the supine
Melody with a simple mineral solution. Within minutes

she hauled herself back onto her hooves and began placidly munching on hay, apparently unaffected by her close
brush with death.

"You look like you could use some help around here,"
said the vet, glancing about with an expression of con-
dolence. "There was a young lad up to my office looking
for work just the other day. He wanted to sweep up
around the place, take care of the animals, that sort of
thing. I had nothing for him, but he seemed like a decent
chap. A Cape Bretoner, I think. Came from a good, frugal
Scottish family, but they've run into hard times."

Phil MacPherson turned out to be a godsend. He
was a robust young man in his early twenties, with blond
hair and white eyelashes that matched those of the Charolais beef critters. Phil and the critters had an affinity
for one another right from the start, and it was thanks
to him that Gordon was able to acquire three more cows
and an electric milking machine.

"I'm going to have to level with you," Phil had said
when he first came to work for us. "I can only stay till
I have enough money saved up to get my father caught
up with his loan payments. Otherwise we'll lose the farm.
Me and my older brother's out working, but my younger
brother's staying home to help Dad with the chores."

"Three boys," said Gordon thoughtfully. "Do you
have any sisters?"

"No sisters."

"And your dairy cows give you mostly heifer calves,
right?"

Phil looked puzzled. "We get our share of heifers,
sure."

"Sons and heifers," laughed Gordon. "Your father's
a good farmer. A real farmer. We're just going to have
to see to it that he gets back on his feet as fast as he can
before the bank takes him over. Because I can tell you

one thing. I've dealt with a few bankers in my time, and you can bet your last dollar that not one of them would be caught dead sitting on a milk stool at five o'clock in the morning."

Phil and Gordon worked side by side from dawn until dusk as new orders for more yogurt came pouring in daily. Sobeys stores at other locations were eager to have the same drawing card enjoyed by the Queen Street supermarket, and the various stores in our vicinity also wanted their share. Once the yogurt found its way into the local outlets, the head offices of IGA, Save Easy, and Dominion were soon on the phone asking for product in the metropolitan area. Even Norman Newman, owner of the Capitol Food Stores in Halifax and Dartmouth, showed up at the farm to see the mysterious new dairy that had made such an unexpected appearance. Like David Sobey, he was surprised and amused to see so much activity emanating from one small kitchen stove. After satisfying himself that we meant business and were here to stay, he readily agreed to let us take on his new stores at a pace we felt we could manage.

No pace, as far as Phil and Gordon were concerned, was unmanageable. Phil was eager to put in as much overtime as possible so as to hasten the rescue of his family's farm, and Gordon, true to his entrepreneurial nature, was busily orchestrating the more or less haphazard growth of our brave new industry. This ultimately meant teaching Phil everything we had learned about yogurt. At first Phil thought yogurt was something people did while standing on their heads. He then convinced himself that yogurt was just another name for the spoiled milk his mother had forced him to drink as a child. When Gordon finally persuaded him to try some of our strawberry yogurt, his face lit up with an expression of delighted astonishment.

"It's like a dish of fresh strawberries and cream!"

he declared. "*Everybody* should be eating this. It's even better than *ice* cream!"

From that moment on, Phil threw himself into the task of producing yogurt with the dedication of a true connoisseur of fine foods. As time went by, he began to develop a scientific interest in the activities of the bacteria responsible for the results of his repeated efforts in the kitchen.

"I'm going to slop this to the pigs and make starter all over again," he announced one morning. "The sour bugs have got ahead of the sticky ones, and I'm getting a lemony product here. I just don't think it's good enough."

Phil had learned that the "sour bugs" (*Streptococcus thermophilus* bacteria) flourished at higher temperatures and were responsible for making the yogurt tart and firm. These had to be carefully balanced with what he referred to as the "sticky bugs" (*Lactobacillus bulgaricus*), whose job was to make the yogurt mild and slightly gluey. These preferred the cooler temperature range during incubation. Phil's goal was to make sure that the two strains of bacteria proliferated in the proper numbers, and this he did by manipulating both the duration and the temperature of incubation.

"It's a real trick to make this here yogurt come out right," Phil went on. "But when it's perfect, I don't mind saying I feel pretty proud of myself. I feel kind of like a craftsman." He thought for a while. "Or maybe even an *artist!*"

A penny saved

If any real planning was carried out in those early stages of our development, it was accomplished on a purely subliminal level, without our really being aware of it. Our growth, while not altogether uncontrolled, seemed to move along quite nicely on its own. Every time we needed more milk to meet the increasing demand, we simply went out and bought another cow. When at last we outgrew our station wagon, we acquired a three-quarter-ton pickup truck for making deliveries, and a supply of new styrofoam boxes was always available at Canadian Tire.

Once Phil was on the payroll, I began to appreciate the enormous importance not only of growth but of cost consciousness as well. It slowly dawned on me that a penny saved is not just a penny earned, but something considerably closer to nine pennies earned, for no more than about 8 to 10 percent of our sales was ending up back in the business again. One of our greatest costs seemed to be the yogurt that would unaccountably remain on the dairy-case shelves unsold until it eventually expired. It turned out that the eagle-eyed customers would buy only the freshest yogurt on the shelf, even if it was only two or three days fresher, so the older yogurt would remain where it was until we finally had to replace it with newly made product. We tried to outfox the customers by cleverly placing the new yogurt behind the older tubs, but nobody was in the least bit fooled by this ploy. The shoppers would simply reach behind the first containers and take the yogurt from the back of the case, leaving the display in a state of total disarray.

Phil and I put our heads together and finally came up with a very complex but worthwhile system of inventory control, whereby we calculated the exact number of individual containers sold in a week and rewarded the stores accordingly. Our system worked quite well. We cut our returns down from about 20 percent to somewhat less than 3 percent, and the few expired yogurts we did find lurking about on the shelves were fed the very same evening to our ever appreciative pigs.

One of our most disconcerting costs was the fuel bill, which had more than tripled since the purchase of our trusty three-quarter-ton pickup. Although I was now getting a free ride in the truck to Dalhousie (where I would teach my classes while Phil delivered the yogurt to the Halifax stores), I still didn't feel that this personal bonus fully justified our greatly elevated gasoline bills. My fear of being drowned in gas-related debt urged me on to greater growth in the metropolitan area so that the pickup could be fully loaded every time it made a trip. I also arranged to buy some of our ingredients at a lower price from a jobber in Halifax so that Phil wouldn't have to drive an empty truck back to Lunenburg.

One of the most obvious ways to cut down on our fuel consumption, however, was to land as many accounts as we could within the closest possible proximity to Lunenburg. The only store in the area not yet carrying our yogurt was a well-appointed supermarket owned by Goliath Food Limited and located in a community called Off Centre (just south of Back Centre and slightly to the left of Front Centre). People would drive from miles around to shop for their aggressively advertised bargains.

"We'd be delighted to take your yogurt," said the manager of the Goliath store. He was a tense, hawk-nosed young man in his early thirties. "All's you do is you make an appointment to see Ed Shoemaker up at head office in Halifax. He's the buyer for Nova Scotia."

He gave me a nervous smile and dashed away, darting furtive glances over his shoulder as though fearful that Mr. Shoemaker might be in close pursuit. I found him a few minutes later pricing a case of baked beans.

"I was just wondering," I began.

"Yes?" he said flatly, compressing his already thin lips into an even thinner line.

"Isn't Goliath's head office somewhere in Ontario?"

"Yes, but you won't be needing to go up there. You just talk to Ed Shoemaker. He makes all the decisions for the Maritime stores."

He continued pricing the baked beans as though his very job depended on his successful completion of the task.

Weak links

After making me cool my heels for well over an hour in the waiting room, Mr. Edward Shoemaker, head buyer for Goliath Food Limited, deigned to summon me. His invitation to join him in the inner sanctum was perfunctorily extended by the receptionist, a nondescript young woman who seemed to be laboring under an intolerable burden of unmitigated boredom.

I let myself into his office and stood patiently in front of his desk, but it was apparently against Mr. Shoemaker's principles to look up when a salesman entered the room. He continued to scrutinize the file folders spread out before him, intent on creating the impression that he was a busy man grappling with urgent matters that needed his immediate attention. His rather low brow was furrowed with the intense effort required to conjure up this image of importance, and his lips moved imperceptibly as he struggled to absorb some of the words appearing in the reports on his desk. Mr. Shoemaker licked his short, stubby index finger and tried unsuccessfully to turn a page.

"Well?" he barked, looking up at me irritably. "I haven't got all day."

I was so surprised by this sudden opportunity to speak that I almost forgot what I was going to say. I cast around quickly for an opener, hoping meanwhile to soften him with a pleasant smile, but he was already leafing through the folders again.

"This, Mr. Shoemaker, is your lucky day," I began.

"I bet," he growled, without looking up. "What's the pitch?"

"There *is* no pitch. I've simply come to tell you that

we're ready now to supply Goliath with the world's best yogurt."

"We have enough yogurt," he said curtly. "We don't need any new brands."

I was dazed. "But you don't understand. It's the best yogurt anywhere! I make it myself, I should know. Here, I brought you a sample of our strawberry yogurt. Have you ever tried it?"

I placed the container on his desk and stood back to savor the full effect of its impact on him. It was a beautifully printed container, featuring Daisy the Cow standing peacefully under a sign that read PENINSULA FARM NATURAL STRAWBERRY YOGURT. The graphic artist had depicted a bucolic farm scene that even a hard-nosed buyer like Mr. Shoemaker would surely find hard to resist. I took comfort in the thought that he, like most people in Nova Scotia, had probably been brought up on a farm. No doubt the sight of the container would transport him back to those happy days, and his attitude toward both me and my yogurt would certainly mellow.

"I don't eat yogurt," said Mr. Shoemaker with a touch of pride. "As far as I'm concerned, it's just another name for sour milk."

My heart was beginning to beat faster. "But why don't you just try some of mine," I suggested, pressing a plastic spoon into his meaty fist. "You won't believe your taste buds. This experience is going to change your life. There's only one drawback, though. Once you taste it, you'll become an addict. Then you'll need your daily yogurt fix, and you'll blame *me* for getting you hooked."

Mr. Shoemaker was unamused. "You couldn't pay me to eat that stuff," he declared, pushing the plastic spoon and the unfairly maligned yogurt back to my side of the desk. I felt myself growing contentious. I knew just by looking at him that he had never tried yogurt before in his life and had no intention of doing so. What galled me was the arrogant self-confidence in which he so easily couched his pitiful ignorance. I felt like wringing his unappealingly begrizzled neck, but instead I merely asked him to think for a moment about the welfare of the shoppers. If he decided to stock our yogurt on his shelves, didn't he think his customers would be delighted to find such an excellent product in the Goliath outlets, and didn't he realize how impressed they would be with the wisdom and discernment of the head buyer for the Maritime provinces?

"The customers are happy with what they've got," he said blandly.

"That's impossible!" I exclaimed, trying to control the urge to open the yogurt container and dump the contents onto his greasy head. A sense of humor in this situation would have been helpful, but I was growing more and more incensed at the attitude of my Neanderthal interlocutor.

"What kind of a deal have you got for me, anyway?" Mr. Shoemaker was saying, inspecting me shrewdly.

"Deal?"

"C'mon, c'mon," he said sharply. "I haven't got any more time to waste on this. If I put your yogurt in my stores, what's in it for me?"

I began to brighten. "We do everything from milking the cows to stocking the shelves in the stores. We also distribute the yogurt in our own truck, and we price it when we put it in the dairy case. We have a good inventory-control system, too. We rotate the yogurt and we credit you for anything that expires. All you have to do is ring up the sales."

"The deal, the deal," said Mr. Shoemaker, rubbing his thumbs against his fingers in a gesture indicating, I assumed, some sort of exchange of money.

"Our suggested retail price gives you a twenty percent markup in the stores," I faltered, wondering what exactly he was driving at. "That's par for the course."

"What about head office?" he demanded. "What kind of a rebate system have you got for us *here?* I'm not interested in your store-level policy. I want to know how much we've got coming to us here at head office."

"I didn't know I was supposed to send money to your head office."

"Look, do you want to get into Goliath or don't you?" he said impatiently.

"Of course I do."

"Then you'll have to raise the ante by another twenty percent if you're going to meet the competition."

"*What?* Are you asking me to send your head office twenty percent of my overall sales, right off the *top?*"

"You got it," said Mr. Shoemaker, with no hint of discomfort.

My initial anger was turning to shock. "I couldn't possibly send you an extra twenty percent. I don't even get that myself! I'm lucky if I end up with ten percent, and that has to be plowed back into the business, anyway. I need that ten percent for growth!"

"Then you'll just have to get more efficient."

Shock turned to fury. "I *am* efficient! Nobody could be any more efficient than we are. We don't waste anything. We get all our cardboard delivery cartons from the other dairies, who leave them lying around in the stores. And we've got our returns down to almost nothing, and whenever yogurt does expire we feed it to the pigs, who then get sold as pork. What other dairy does that? And what's more, we don't have any middleman costs because we do it all ourselves. There isn't one single area of our business that could be run any more efficiently!"

"Then get bigger."

I looked at him numbly. "How can I, when you control most of the stores in the Maritimes?"

"Just raise your price by twenty percent, then. Get it out of the customers' hides."

"I can't do that," I sputtered. "If I raise my price by twenty percent, I'd probably lose half my sales! What good would that do?"

"I thought you said you made the best yogurt in the world," he smirked. "If that's the case, then the customers will pay whatever you want them to pay."

"That's not why I make the world's best yogurt, to *gouge* the customers," I said indignantly. "I want to sell them an excellent product at a fair price."

"Just what makes you really think your product is so excellent, anyway?" he asked, leaning back and folding his arms.

"Well, we use fresh milk, for one thing. And we don't cook our berries. They're fresh frozen, and the aroma is . . ."

"There's your problem right there," he interrupted. "You're going to have to cut your costs, then, if you want to be competitive. Don't use fresh milk. Use milk solids, like everybody else. And get rid of your damn frozen berries. All the other dairies use jam, or some kind of a cheap preparation like that."

"But I don't want to make a cheap yogurt," I objected.

"Look, Mrs. Jones, you're running out of options. If you want to make a fancy specialty item, then stay small and keep your prices high. People are always saying that small is beautiful. So prove it. But when you get tired of milking cows and making yogurt in your kitchen and dipping the stuff into containers by hand, come back and we'll have another little talk. Maybe by then you'll be ready to see things *my* way."

He got up and led me abruptly to the door. I had been dismissed. The receptionist never looked up as I stalked out of the Halifax headquarters of Goliath Food Limited, a company whose shelves, I vowed, would someday be creaking with massive quantities of Peninsula Farm yogurt. And at the right price.

8

Eat It, It's Ecological

Managing the cornucopia

Ginger Corcoran arrived at the kitchen door just as the sun was setting on a perfect June day. Her hair was almost the same color as the spreading sunset, and her pale skin was sprinkled liberally with freckles and small specks of mud. She wore blue shorts, a faded T-shirt, and an ancient pair of Adidas sneakers. She smelled faintly of a combination of peppermint and damp earth.

"I'm from Corcoran's Nursery," she said, extending an eager hand. "I'm selling seeds and seedlings, and I have all kinds of anything you want out in the car. Have you got your garden planned for this summer?"

She looked like a seedling herself, I thought, as I contemplated the young girl framed in the doorway.

"As a matter of fact, we haven't planted a garden."

"Oh, that's terrific," said Ginger, breathing a sigh of relief. "I've been peddling seeds for the whole month of May, but most people already have their gardens planted now, and I haven't been getting any more customers for the last few days."

She stood there smiling hopefully at me while I struggled with the decision as to whether or not to put in a garden that summer. The thought of fresh vegetables on the dinner table was certainly tempting, but I was not enough of a masochist to take on the extra work of tending a garden. This I communicated to Ginger as tactfully as I could, explaining to her that our budding yogurt business was taking up all our free time, but she was not about to take no for an answer.

"Let me work for you, then," she pleaded. "I could take care of the garden, and I'm *great* with children. I could learn to make the yogurt, if you'd teach me how. I could even deliver it for you, too, if you'd let me use your truck. I'm a fast learner. Really I am."

"It's not that easy to make yogurt, Ginger," I said gently. "It's sort of like making wine, when you get right down to it. You have to be terribly accurate with times and temperatures, and the yogurt has different vintages, depending on how successful you are. It's tricky, believe me."

She believed me, but she remained undeterred. She was so keen to come to work for us that we finally agreed to put her on the payroll the very next week. She arrived in the kitchen all scrubbed and shiny, with her unruly red hair stuffed carefully into an ample hairnet. She was wearing a white nurse's uniform of which she was inordinately proud.

"It makes me feel like a real professional," she said happily. "Now the yogurt will just *have* to turn out right."

Ginger was as good as her word. She absorbed everything we taught her as quickly as could be expected, but she was so eager to progress that she didn't always tell us when she had not fully understood what she was learning. The pigs didn't mind. They gobbled up her mistakes with their usual élan, but the financial strain of having a trainee on board did not go unnoticed by the payroll department working out of the spare bedroom upstairs.

"She's overlapping with Phil," said Gordon, examining our monthly ledger with some consternation. "It can't be helped, of course, but until she can work on her own she's going to cost us a bundle. We'll just have to increase our sales a little faster."

By this time we had begun to sell off our beef critters, replacing them with dairy cows as the need arose. We had acquired a grand total of twelve Jerseys, but we calculated we needed at least three more to cover the cost of carrying Ginger. When all fifteen of them went out on the grass for the summer, we discovered to our surprise that they were providing us with far more cream than we could use. Excess milk was one thing, but the thought of slopping beautiful, rich, yellow Jersey cream to the pigs was more than we could bear. Gordon sought help from the product development department.

"You'll have to come up with a recipe for ice cream, Sonia," he announced one morning. "We can't waste all that cream. And besides, the chickens are laying more eggs than you can count. Now that Ginger has them eating the garden slugs, they seem to think that gives them carte blanche to raise their egg production. We're going to have to do something to manage this cornucopia of farm products."

When Ginger wasn't using the stove to pasteurize milk for yogurt, I would slip into the kitchen and work feverishly to develop a custard to use as a base for the ice cream. I discovered that custard works out perfectly when it is made in the quantity described in a normal cookbook, but when I tried to raise the batch size from six eggs to thirty-six eggs, I found that the custard had an annoying habit of either turning to scrambled eggs or flatly refusing to thicken at all. The pigs had no complaints whatsoever. If anything, they liked curdled custard even more than expired yogurt, so I was able to keep them in ecstasy for days on end while I struggled to master the elusive art of turning ordinary brown eggs into creamy, lumpless custard.

I grew to resent those eggs. They seemed to take real pleasure in scrambling on me, as though they were getting back at me for stealing them away from the hens. I eventually learned that I could foil the eggs by keeping a candy thermometer in the custard and whisking the mixture off the burner at exactly 176 degrees Fahrenheit. I would then plunge the whole pot into a sinkful of cold water before the eggs had a chance to become lumpy. Only occasionally would I allow myself the pleasure of a small chortle of triumph.

The next step was to develop the ice-cream flavors. I soon discovered that slightly scalded cream tastes even more delicious than lightly pasteurized cream, and when mixed with egg custard, sugar, and pure vanilla the results were enough to please even the most skeptical taster. The French vanilla mixture became the basic batter for a host of other flavors that came to life almost as quickly as I could think them up. Cocoa and chocolate syrup made a Dutch chocolate ice cream capable of satisfying any palate, and a slurry made of instant coffee and hot water turned the basic batter into a café au lait whose strength could easily be varied as desired. It turned out that the sum of these two flavors of ice cream was even better than their parts, for when I mixed them both together I had a mocha ice cream that was hard for any of us to resist.

The experiments went on and on: maple walnut made from pure Nova Scotia maple syrup; strawberry and raspberry made from our own supply of fresh frozen berries; lemon and lime made from frozen concentrated juice; orange ice cream mixed with hand-shaved chocolate; Black Forest ice cream made from the chocolate batter blended with cherry brandy and brandied cherries; Rum and Raisin made with pure Bacardi rum. The list could have been almost interminable, but I thought it prudent to limit ourselves to about ten flavors or so, at least until we found out what kind of demand we could generate for this hand-cranked, homemade ice cream.

Peninsula Farm becomes a tourist attraction

As usual, we were not prepared for what ensued. Visitors came flocking to the farm in droves, drawn to us by a little brochure we had deposited in the Lunenburg tourist bureau. The pamphlet featured Daisy the Cow and a short history of her contribution to the growth of our company, and it also provided a brief description of the various flavors of yogurt and ice cream we produced from her milk. Soon there was a line of people standing patiently by the kitchen door waiting for ice-cream cones. Inside the kitchen Ginger continued to make yogurt as I frantically dashed around brandishing an ice-cream scoop in one hand while trying to make change with the other.

Enter Edna, a portly, middle-aged housewife from down the road who took pity on us one day and offered her services as an apprentice ice-cream maker. We had no choice. It was either Edna or the mental hospital. Edna was eager to be of help, but her training period was lengthened substantially by her firm belief that after thirty-five years in the kitchen she did not need to be taught how to make custard. The eggs immediately took advantage of the situation by turning themselves into tender lumps. The pigs were treated on numerous occasions to their favorite dessert of scrambled-egg custard until I was at last able to convince Edna of the supreme importance of the candy thermometer. When she finally accepted the fact that her experience had to be modified to fit the new circumstances, she became a model ice-cream maker. Her unflappable temperament was ideally suited to the usual uproar in the kitchen, and she calmly churned out gallon after gallon of delectable ice cream while I tried desperately to cope with the growing numbers of tourists.

Ginger, meanwhile, was forever demanding new styrofoam boxes in which to incubate the constantly increasing quantities of milk that Gordon left at the kitchen doorstep. These she stacked neatly against the walls in the halls and living room, and some even ended up in the dining room. Our children grew up taking it perfectly for granted that a house was like a large igloo lined with styrofoam boxes. At first I was worried that the little girls would somehow get lost in the shuffle, but they were secure in the knowledge that they could always find Gordon or me in either the house or the barn, and they would toddle back and forth checking up on us to make sure everything was running smoothly. When they found things running too smoothly, they would toss an occasional monkey wrench into the works, but the business managed to survive in spite of everything.

Flourished is perhaps a better word for it. This was due in part to the tourists, who had ideas of their own as to what should be awaiting them on their arrival at the farm. First of all they were disgruntled by the absence of coffee and sandwiches to accompany their ice cream. They didn't want to have to turn around and go to town for lunch and then come all the way back again for dessert, so I agreed to make them egg-salad sandwiches from the extra eggs provided by our hens. A one-item menu, however, seemed somewhat limited, so I decided to add ham and pork sandwiches to the list of selections. We all felt sincerely saddened by the obvious inequities of the donations made by the chickens versus the pigs, but there was really no way we could allow the pigs to live on forever. We tried to remember them as having enjoyed a short but happy life, filled with scrambled eggs and yogurt. We also consoled ourselves with the thought that at least the customers had never known them by name. Their joy upon biting into the canapés made with homemade pork liver pâté mixed with chopped eggs and Courvoisier helped to ease us over the sorrow we felt at the loss, every five months, of our latest batch of pigs.

As for the tourists, they contributed more than they realized to the operation of the farm, for every time they took a bite of food they were helping to round off the ecological balance we had so carefully established among the animals themselves. Our costs, in those early days of our development, were greatly reduced by this delicate balance. The chickens would fatten themselves on maggots, slugs, and various garden pests; the pigs would take care of the wastage emanating from the kitchen or the dairy-case shelves; and all the animals would keep the land and the garden well fertilized. Finally, the tourists would liquidate the farm-related food products so that the cash could then be used to keep the business growing, thus completing the circle. The ecology of the farm depended to a large extent on the healthy appetites of both man and beast, and nobody appreciated this more than those of us who were firsthand observers.

Our visitors, however, expected not only a full meal when they came to the farm, but a grand tour of the premises as well. Not that there was anything to see. On a twenty-five-acre farm everything is in full view, and a guide would have felt rather foolish standing in one place and pointing out the farmhouse, the barn, the grazing cows, the electric fences (beware!), and the distant shoreline. But Gordon understood their needs far better than I did. He realized that the vast majority of visitors had never set foot on a farm before. Many of them were from Toronto, Montreal, Boston, or New York, and they were curious to see how rural people lived and what they did from day to day. They were glad to be away from their office jobs in the city, and perhaps some of them were entertaining secret thoughts of eventually retiring to the country. They had stumbled upon this small farm in Nova Scotia, and it was up to Farmer Jones to show and tell.

As the days and weeks passed, Gordon slowly developed an informative lecture that he presented to the tourists as he performed his duties in the barn. He discussed anything and everything from agricultural economics to basic genetics, while the cows peacefully munched away in the background, oblivious to their newly found status as Exhibit A in the growing saga of Peninsula Farm. The visitors loved it. They came away feeling they had learned a great deal about farming in the course of an hour, and they no doubt developed a deeper respect for farmers and maybe even a greater appreciation for the food on their tables. Gordon was happy to be of service, but he didn't have the heart to mention that he was really just a boy from Brooklyn who four years earlier had never dreamed he would end up on a farm in Nova Scotia showing fellow New Yorkers how to milk cows.

A tiger by the tail

It was a typical Sunday in August. Phil was mending fences, Gordon was paying bills, Ginger was inoculating milk, Edna was making ice cream, Valerie was tasting the batter, Vicki was solemnly smearing chocolate syrup on her face, I was making a stack of assorted sandwiches, and the tourists were waiting with saintly patience by the kitchen door. The afternoon progressed with its usual mixture of hard labor, hilarity, and general confusion, until finally I found myself taking care of the last customer of the day. She was standing between the dishwasher and the kitchen table, surveying the pandemonium around her with an expression of discreet horror.

"You can't really be making all your yogurt here in this kitchen!" she exclaimed, looking dubiously at the pots and pans that had accumulated in startling quantities on the floor around the sink.

"Well, we package it in the dining room, and we incubate it in the hall and living room, so we're not just working out of the kitchen."

She stared at me. "But where do you *live?*"

"We live mainly upstairs," I said, almost apologetically. "But we do manage to get downstairs once in a while, when there's room for us to move around."

"You're going to drive yourselves crazy! What will you do when the yogurt takes over the whole house? Where will you go then?"

I tried to change the subject. "Would you like mustard or mayonnaise on your sandwich?"

"Mustard, thank you very much. Look, I don't want you to think I'm out of line for asking you all these questions, but I do think you have a problem, and I may be able to help. You can't go on like this, making yogurt in your kitchen. What you need is a dairy plant! A small dairy plant, of course."

"We couldn't possibly afford to build a dairy plant," I smiled, handing her the sandwich. "We're just barely balancing the sales with our costs and the payroll as it is. A dairy plant would be out of the question. Even a small one."

"Have you ever thought of applying for a government grant?" she asked, taking a tentative bite from the corner of her sandwich.

"A government grant? To put up a dairy plant? But I'm only a housewife working in the kitchen. If government agents ever came here and saw me doing this, they'd die laughing."

"I'm not laughing," she said soberly.

It was my turn to stare. "You're from the government?"

She looked slightly embarrassed. "I'm sorry. I should have told you before, but the truth is I only came here for an ice-cream cone. But then when I saw you all working so hard in the kitchen, and when I realized how desperate you were for some proper space and equipment, well, I thought I'd just mention our granting programs. The problem is, though, we're not really supposed to solicit applications. But this is my day off, and I just *had* to say something. Besides, I have a selfish motive, too. I'm one of your biggest yogurt fans, and I'd just hate to see you get swamped and give up or something like that. This province needs high-quality products. Especially yogurt!"

After we put the children to bed that night, Gordon and I sat down at the kitchen table and tried to come to some sort of an agreement about what direction the business should take. As Gordon saw it, there were only two ways to go: either we did it right, or we threw in the towel.

"We really have no choice," he said, pouring himself a second glass of beer. "We can't go on making yogurt on the kitchen stove. We're doing over a thousand dollars a week in sales now, and I'd say that was pretty good, considering everything. But people are going to keep wanting more yogurt, and we're already at the end of our rope. What we need is a small building with a walk-in cooler and a proper pasteurization vat and a good table to work on. And it wouldn't hurt to have some storage space and a holding tank for the milk, too. Plus we need a walk-in freezer for the fruit and the extra ice cream."

"So how do we pay for it?" I asked nervously.

"No problem. The government contributes twenty percent and we borrow the rest. We own the house and the farm free and clear, so we have plenty of collateral. Then we use our savings for cash flow until we get back on our feet again."

I tried to mask my growing panic with an expression of beatific calm. "And what if we don't get back on our feet again?"

Gordon shrugged. "We'll have to, that's all."

"But isn't it kind of risky? We're staking everything we own on the success of this business!"

"I'd rather stake my money on myself than invest it in somebody else's business. This way at least we're in control of our own destiny. We'll work hard and we'll make our own decisions. We'll be okay, you'll see."

"I do have a job, you know. It isn't as though we *had* to do this. We have a choice. And what about your yacht?"

"The yacht, yes," said Gordon dreamily. "It would be nice just to sail away into the sunset. But it's too soon to be doing that at this point in our lives. We have to contribute something first, and yogurt is as good a contribution as anything I can think of. We'll create more jobs and buy lots of fruit and put a few more cows to work, and a good many people will benefit from the whole operation. And on top of that, it'll be *fun*. It'll be a challenge to see if we can put real food back on the market again. There's too much junk everywhere. It's ridiculous."

"Some of my colleagues are still grumbling about my involvement in the business," I ventured. "They think I ought to be spending a hundred percent of my time at the university."

"Professors love to grumble," Gordon laughed. "Did you tell your supervisor that your textbook has already been adopted by thirty-five universities in its first year of publication? That should keep him happy for a while."

"But he doesn't seem to be very happy. When I told him about the adoptions, all he said was that textbooks don't count as far as he's concerned. He'd like to see me write a scholarly book now. One that would make a significant contribution to the research being done in my field. He wants the people in our department to write material with academic impact, as he calls it. He probably wants to impress everyone with his annual report."

"He sounds like a real schlemiel to me. Who is he going to impress, anyway? He wants you to write a book that'll be read by only three other specialists in the world, and even those three won't read it for the fun of it. They'll read it so they can criticize it and tear it down. What a royal waste of time!"

"I have a topic I want to write about, though. And I have a publisher for it too, so I think I'll go ahead. But don't worry, I'll write it in plain English, with no jargon, so I should be able to double the readership, anyway. Then *six* critics can take pot shots!"

"I hope you know what you're doing," said Gordon, shaking his head. "Life is so darn short, and it's important to make the right choices. It seems a shame to me that you have to spend so much time trying to please that supervisor of yours. I'd much rather have you dream impossible dreams with me. How about it? Do you want to become a yogurt queen and bring pleasure to countless taste buds, or would you rather be an obscure, unread, unappreciated scholar, having little or no influence on the brain cells of the world?"

We spent the rest of the evening grappling with the relative importance of the five senses and their place in our lives. We realized that we had arrived at a crossroads and that a decision had to be made about whether or not to invest everything we owned in a new yogurt factory. It seemed to us that the occasion called for some deeply philosophical comments about the meaning of it all, but our decision to go ahead with the factory was ultimately based on nothing more than some very pragmatic reasoning. We had started what promised to be a very successful little business; we were developing a small following of dedicated yogurt lovers; the stores were depending on us to continue servicing them; and we were providing employment for local cows and assorted residents of the province. We couldn't give up. We had a tiger by the tail, and we were determined to hang on.

9

North America's Smallest Dairy Plant

A giant step backward

Help, we thought, had come just at the right moment, but we were completely wrong about the timing of the construction of our new yogurt factory. The contractor had assured us that the building would be up and ready for us to move in by December, but when Christmas came there was nothing on site but a thirty-by-thirty-foot cement pad surrounded by great mounds of sodden earth.

"Production is being held up at the other end," said the contractor dourly. "The workers who make the steel buildings are on strike, and it looks like it'll be a while before they can ship us the beams and girders and all the rest of the material."

"Well, let's get them from somebody else then," said Gordon, exasperated.

"Can't do that. I've already given them the order, along with your down payment. It's best just to sit tight till they straighten it out. Them guys always get what they ask for. It don't matter what it is, they end up getting what they want." He spat discontentedly. "It sure beats being in business for yourself," he added bitterly.

Spring arrived, and the steel building still hadn't

been shipped. The strikers had long since been granted their demands, but it seemed that the manufacturer was behind schedule and had somehow managed to put us at the end of what must have been a very long waiting list. Meanwhile, life in the kitchen was getting busier by the minute. When summer ended and Edna no longer had to make ice cream, we switched her over to the position of assistant yogurt maker under the direction of either Phil or Ginger, depending on who was working at the time. Our production level had continued to increase throughout the winter, so we made yogurt seven days a week and distributed it every day except Sunday. This was accomplished by a carefully devised rotation system that involved juggling a forty-eight-hour week with the constant demands of employees, store managers, and our ever expanding herd of cows.

As fate would have it, the building manufacturer picked the muddiest day of May to dispatch our long-awaited dairy plant. No sooner did the tractor and flatbed heave their way over the hill than they bogged down up to the axles in the soft ground on our unpaved road. A corpulent truck driver of massive proportions climbed heavily down from the cab, sputtering and cursing the day he had been chosen to make this most undesirable of all possible trips. He stared petulantly at the road and stamped his foot on the ground a few times, as though unwilling to believe it had given way so easily to his twenty-two wheeler.

"The Department of Highways had their boys out on the road today," he complained. "They stopped me as I was driving in from Blockhouse and they fined me for being overweight. Can you believe that?"

I glanced quickly at his double chin.

"They were weighing everything on the road that

wasn't a VW," he went on. "There's no way a truck can get by the spring weight restrictions they've got posted everywhere. How'd they expect me to get here from the highway? *Fly?*"

Just then Gordon drove up with the farm tractor and managed to connect a chain to the tow-hook on the bumper of the disabled vehicle. The stout driver hauled himself back into his cab with unexpected agility and cautiously touched the accelerator a few times in an effort to help Gordon dislodge him. The two men finally succeeded in freeing the entire rig, which then proceeded uneventfully to the construction site. The driver and his passenger unloaded the steel building by unchaining the load and skillfully driving the flatbed out from under it.

It took the contractor and his men only three days to erect the steel building, but unhappily their admirable speediness turned out not to be a foreshadowing of things to come. Special walls made of an amalgam of fiberglass and plastic had to be ordered from New Hampshire, whereupon they got held up for several weeks at the border while various officials decided what to do with them. Meanwhile, some floor tiling was ordered from a company in a distant province, but the order was lost not twice but three times before the wrong tiles finally arrived. When these were sent back, the right tiles were duly forwarded along with the wrong grout. When the company was informed of the error, the head office advised us that they manufactured only the wrong grout, but recommended that we apply to one of their competitors for the right grout. This turned out to be impossible, as the competitor had been put out of business the year before by the very company with whom we had been corresponding. We finally located the right grout in a hardware store in Left Centre, but they had too little in stock for us to complete the job, nor did they seem to know who their supplier was. Gordon agreed under great duress to let the contractor do the job with a mixture of the right grout and the wrong grout.

The same story, with very few variations, was repeated when it came to installing the ceiling, putting in the sewage disposal bed, and locating rustproof accessories of various descriptions so that the whole plant could be hosed down at the end of each day. But the worst comedy of errors occurred when we greenhorns put ourselves at the mercy of dairy equipment manufacturers and their sales representatives. Since our cows were at that time producing about sixty gallons of milk a day, we decided to order two one-hundred-gallon jacketed vats in which to pasteurize, cool, and mix the yogurt. Our first problem concerned the motors that turned the agitators. These arrived wired for three-phase power, but since we lived on a small farm and not in an industrial park, we had no access to three-phase power. It was difficult for the manufacturer to grasp this essential fact, because *all* normal, sensible dairy plants use three-phase power, but they reluctantly agreed to take back the motors and replace them with single-phase units. These cost us almost twice as much as the original motors, but at least the difficulty was solved.

Our second problem, however, was of a much more serious nature. The agitators in the vats were shaped like long upside-down Ts and tilted on an angle. They did an excellent job of stirring the milk while it was being heated and cooled, but when it came to mixing yogurt, they were absolutely useless. Sixty gallons of beautiful yogurt remained resolutely in place while the T simply cut through it somewhere near the bottom of the vat. There was nothing for it but to scoop the yogurt out with a ladle and mix it with the fruit by hand in a stainless steel pot. We were not much better off than when we had been working in the kitchen.

"What we really need," I explained to the sales representative, "is an agitator shaped like the baffle in my little household ice-cream churn. You know, something in the form of an H that would scrape the sides of the vat. Could you please exchange the agitators you sent us for the right ones?"

No, that would be impossible. The agitators that came with the vats were tilted, but scrape-surface agitators would have to be mounted straight up and down. In order to install the right agitators the entire top of the vat would have to be cut off and rewelded. It would be cheaper to buy new vats.

I felt my heart pound faster. The sales representative was supposed to be an "expert" in the field and was the president of a consulting firm that specialized in supplying the dairy industry with manufacturing equipment. This expert had supervised the development of dozens of dairies across Canada and had come to our farm bearing a briefcase bulging with colored photographs of his sparkling successes. Such an experienced and well-informed individual could not be wrong. There must be a good reason why our vats came with agitators that wouldn't mix yogurt, and the consultant would be eager to put matters right at the earliest opportunity. All I had to do was proceed calmly.

"Well, since you sent us the wrong agitators, I'm sure you'll arrange to have the vats exchanged as soon as possible," I said, trying my best to hide my anxiety.

"I can certainly arrange to have them exchanged," he said smoothly, "but I'll have to charge you for the new vats, of course."

"But why?" My voice had switched to a higher register.

"I sent you just what you needed. You said you wanted pasteurizing vats. You didn't mention you were going to use them for mixing yogurt."

I began to panic. "Yes I did! I remember distinctly telling you that we were going to have to add gelatin to our product because since we were setting it up in sixty-gallon batches we had no way to cool it, and if you disturb warm yogurt to mix in the fruit, it turns to a milk shake and never sets again. And you said nobody minds gelatin, it's a natural product and not a chemical additive. Don't you remember? You *knew* we had to mix the yogurt and the fruit in the vats. There was no other way. You said so yourself!"

"Listen, don't get excited. I'll look into it, okay?"

I had no choice but to let him look into it. We had already paid for the vats in full.

Our third problem threatened to drive us certifiably insane. The filling machine, which came highly recommended by our friend the expert, arrived miraculously on schedule. We had made the usual financial arrangements of one-third down, one-third when shipped, and one-third on delivery. Gordon had checked into the reputation of the manufacturers and found them to have impeccable credentials. So it was with great expectations that we all gathered around the filling machine for its maiden performance. It worked perfectly during the dry run, dropping our printed bowls into the filling slots with admirable precision, then snapping the lids on the bowls with an accuracy that held us all in thrall. But when it came time to fill the containers with actual yogurt instead of air, the filling machine promptly went berserk. It leaked yogurt from every gasket, crack, and orifice, covering itself, the containers, and the floor with smears and streams of luscious strawberry yogurt.

Charlie Chaplin could not have improved on the scene that followed. Edna and Ginger grabbed some paper towels and frantically wiped the containers as they emerged from the machine; Gordon began fiddling with knobs and levers; I did my utmost to keep the packing table free of the ensuing yogurt container traffic jam; and Phil decided his best contribution was to hose down everything in sight, including us.

Once the hopper was empty and we had a chance to collect our thoughts, I picked up the telephone and dialed the sales representative in hopes of receiving some expert advice. The operator cut in.

"I'm sorry, that number has been disconnected."

"But that's impossible, operator. I'm calling a business number. There must be some mistake."

"I'm sorry, Madam, but that number is not in working order."

With pounding heart and trembling fingers, I dialed the manufacturer and proceeded to pour out my woes to a very sympathetic service manager.

"What did you say the model number was on that machine?" he asked, when I finally came to the end of my litany.

"It's a model five hundred."

"No wonder it's leaking, then. The model five hundred was never meant to fill yogurt. It was designed for filling cottage cheese, or something of more or less the same viscosity."

I took a deep breath. "But your sales rep specifically recommended that model to us. He said it would be *perfect* for filling yogurt!"

"I'm not surprised. He happened to have that particular machine on his hands, and he managed to unload it on you just before he went out of business."

My worst fears were confirmed. "Can you take it back, then, and exchange it for a *yogurt*-filling machine?"

"I'm sure we could make some kind of an arrangement. Delivery time on the new machine would be about six months."

"Six months?" I cried. "But we can't wait that long! We're already drowning in yogurt!"

"Have you ever considered making cottage cheese?"

"I don't want to produce cottage cheese. There's a local dairy here that already makes it very well, so we wouldn't want to hurt them by doing the same thing. What we need up here is yogurt, but now I don't know how we're going to be able to get it packaged and out on the market."

"We've both got our problems, I guess," said the service manager morosely. "Our sales rep skipped town with your payment, so it looks like we've both been left holding the bag. Anyway, if you want that other filling machine, just let me know. We wouldn't be able to give you much on your old one, though. A machine like that depreciates about fifty percent the minute you press the start-up switch."

Inedible mistakes

Although we had come to rely heavily on the pigs to eat our mistakes, the poor creatures could do nothing to help us dispose of the stainless steel vats and the model 500 filling machine. It was up to us to figure out new ways of turning our many false steps to our own advantage.

"Don't worry, we'll come up with something," said Gordon confidently as I eyed the mounting pile of bills on his desk. "Mistakes are just another way of doing things, you know. They'll force us to be more creative, and that's half the fun."

The problem of the mixing vats caused creativity to strike the very next day when Gordon came home with a Waukesha pump and a length of clear plastic hose. First, he attached one end of the hose to the outlet on the bottom of the vat; then he draped the other end over the edge of the top, holding it in place with the stainless steel lid. The idea was to circulate the yogurt in and out of the vat through the hose and pump until the entire batch was thoroughly mixed, but once again we managed to fall victim to a small oversight. We had neglected to take into consideration the pressure of the pump, so when Gordon flipped the control switch the hose under the lid jumped off the vat and proceeded to dance wildly all over the factory, gleefully spreading yogurt in every direction. Gordon dashed around after the hose, slipping and sliding on the growing puddles of yogurt on the tile floor, while I watched stupidly from a safe distance. It took me several seconds to realize that the solution to our present crisis was simply to turn off the pump, but by that time Gordon had already landed the recalcitrant hose with a flying tackle.

It turned out, however, that our secondhand Waukesha pump (purchased impulsively without benefit of expert intervention from the consulting sector) allowed us to get the better of the mixing vats that wouldn't mix. Gordon marched off to the county jail in search of a pair of handcuffs to use on the unruly hose, but finding none available he reluctantly settled for a piece of nylon cord with which he made a hangman's noose to secure the neck of the hose over the lip of the vat. Once this was accomplished the yogurt circulated calmly while Phil poured in the appropriate amounts of fruit and sugar. The results were excellent. By that time we had developed recipes for maple, peach, and raspberry yogurt to add to our original repertoire of plain and strawberry, and the pump did a superb job of blending these flavors into a smooth and colorful mixture.

The filling machine, on the other hand, continued to do everything in its power to obstruct the packaging process. After weeks of frustration, we were able to come to terms with the drips and leaks by the judicious use of squeegees, suction pumps, washdowns, paper towels, and strategically placed buckets. But the lidder was not brought so easily under control. Either it would use too little pressure in placing the lids on the containers (leaving Ginger the task of snapping them on by hand), or it would decide to jam the lids on so hard that the containers would burst open and shoot their contents all over Edna. After a few weeks of this treatment, Edna learned to detect a certain warning click in the machinery that allowed her to adjust the pressure just in time to avoid being given a yogurt bath, a feat that raised her spirits considerably. She was a good sport about it all, though, cheerfully claiming that her splashdowns had taught her that yogurt facials were wonderful for discouraging wrinkles.

"Do you realize how much money we're losing on this filling machine?" said Gordon one afternoon as he glowered at the three slop buckets brimming with yogurt spills. "Can't you invent a new dairy product that doesn't drip? If we could use this filling machine to package a product that can't leak out of it, then we could justify buying a new one that would work properly for yogurt. God, if we had the right equipment, the sky would be the limit!"

I found my new assignment both challenging and intriguing. I loved the idea of developing a brand-new dairy product that had never been on the market before, but my imagination didn't seem capable of rising to the task of dreaming up something of unparalleled originality, for I drew a complete blank every time I tried to wrestle with the problem.

As so often happens in these cases, the solution came to me when I least expected it. Several days earlier I had sat down to enjoy a large container of plain whole-milk yogurt. I had just devoured two or three spoonfuls of it when I was interrupted by a minor crisis and had to replace the bowl in the refrigerator. The next day the spoon holes had filled up with whey (a clear, light yellow liquid natural to cultured milk products), so I poured it out and began to eat the remaining yogurt. This process was repeated for three days in a row as crises, both major and minor, ganged up on me for the express purpose of preventing me from ever finishing the entire serving.

My interest in that particular container of yogurt increased with every passing day. The more whey I poured from the spoon holes the thicker the remaining yogurt became, until finally it reached the consistency of a smooth, creamy spread. It was a little thicker than sour cream but not nearly as stiff as cream cheese, and it obviously had far fewer calories than either. My mind began to dance with possibilities. I could mix it with herbs and onions for a chip dip. It could be blended with olives, or smoked salmon, or caviar for canapés. The base could be used for cheesecakes, or it could be thinned down with milk for a low-calorie salad dressing. It could be plopped on top of a baked potato or blended with fruit and cream for a high-calorie dessert. There was no end to it. Now all that remained was for me to figure out how to make it in large quantities using a procedure more suitable for industrial purposes than my spoon-hole method.

I experimented with every drainage process I could think of, including strainers, punctured trays, and a few methods that were such failures I can't even remember what they were. The pigs grew fat on my false hypotheses. I finally concluded that the physics of the situation required that a certain mass of yogurt be subjected to the earth's gravitational pull by hanging in an unreinforced container whose apertures were large enough to let the whey out but small enough to keep the yogurt in. I had just rediscovered the cheesecloth bag.

"Delicious!" exclaimed Gordon, holding up a Wheat Thin dipped in a blend of drained yogurt, cream, thyme, onions, and salt. "I could eat a ton of that stuff. What are you going to call it?"

I thought for a moment. "I guess I'll call it yogurt cheese. It's a soft, unripened cheese made from yogurt, so that's really the best name for it."

"It's the only name for it," Gordon agreed. "How much of it did you make, anyway?"

"About three gallons."

"Three gallons?" he echoed, jumping up from his chair. "Is anyone using the filling machine right now?"

"No. Edna's just cleaning up, and Ginger went to the barn to help Phil with the cows."

"Then what are we waiting for?" he cried, flinging open the kitchen door and striding toward the factory.

"Daddy, wait for us!" chorused the little girls as they ran to catch up with him.

Soon we were all gathered around the filling machine watching enraptured as it filled and lidded some white sample containers with our new yogurt cheese. We couldn't believe our eyes. The machine was on its best behavior, filling all the containers with the exact amount and marching them off to the packaging table without so much as a smudge of spilled product anywhere in sight. An aura of veritable bliss seemed to emanate from the quietly purring machine, as though it were profoundly satisfied to be fulfilling the purpose for which it had originally been designed. It had found its true calling at last, and nobody could have been happier than those of us who were gathered around it on that memorable afternoon.

10

Surviving

Round pegs, square holes

Once the start-up difficulties in the dairy plant were brought under relative control, we all had to agree that our new machinery had greatly facilitated the production of the yogurt. By June the plant workers had settled into a comfortable routine, finishing by noon what would have taken them all day to accomplish in the kitchen.

"Medium is beautiful, small is a drag," Gordon pronounced, looking with satisfaction at the gleaming stainless steel equipment standing on the freshly washed red tile floor. "We're a long way from the kitchen now!"

"Maybe so, but all this has to be paid for," I said, hesitating to bring up such a mundane topic as money when Gordon was so obviously glorying in the sight of our gleaming new factory. For splendid as the factory was, the final cost was well over twice the original estimate, bringing the government contribution down to below 10 percent. I had never before been so deeply in debt.

"Don't forget, we have the new truck box to pay for, along with the shelves and the refrigeration unit, and there's the incubator, too. Plus the storage charges on the berries in the freezer locker. I never knew we'd have to buy a whole *year's* worth all at one time. Do you realize we could rent a perfectly decent apartment in New York for what it costs us to house those idiotic berries?"

"Don't react to it emotionally. Just think of it in terms of cold, hard numbers," said Gordon, unperturbed by the idea of having to deal with a few extra zeros here and there.

I was given no sympathy from anyone except an itinerant fruit salesman who happened to wander into the plant one day, resplendent in pinstripes and paisley. He was convinced that the exorbitant cost of using fresh frozen fruit in our yogurt would soon turn us belly up. But we were not to worry, for help was on the way. His company could provide us with aseptically packaged fruit that not only preserved the full flavor of the vine-ripened originals, but could be stored indefinitely at room temperature. He promised to send us a sample of this miracle of modern technology, and he defied any of us to distinguish between his product and nature's own.

He was true to his word. His sample arrived several days later in an enormous clear plastic bag about the size of a pillowcase, bulging with some material of indefinable color described on the label as peaches. Phil wrestled the thing into a clean vat, and Gordon stabbed it smartly with a pointed knife while Ginger and Edna stood by. A viscous liquid oozed forth, filling the air with a fragrant odor that was vaguely reminiscent of martinis and eau de Cologne. We all plunged our spoons into this dubious substance, but not one of us could have identified it as peaches had we not been so advised by the label on the plastic bag. Only the pigs displayed any measurable amount of enthusiasm for the product.

We had only just recovered from this experience

when still another fruit salesman appeared on the scene, offering even greater assurances of flavor and quality than the aseptic peaches man. His strawberries, he told us fervently, were in every way the equivalent of fresh berries in the field, and thanks to his company's fastidious heating and cooking methods, they could also be stored for several months at room temperature. Hoping to impress us even further with the excellent properties of his strawberries, he casually mentioned that one of the national yogurt manufacturers used his fruit in their product. He no doubt took it for granted that we were struggling desperately to match their level of achievement.

The sample arrived in a thirty-pound pail with a list of ingredients pasted prominently on the lid. It seemed strange to us that a bucket of strawberries should have a list of ingredients at all, and we were even more puzzled to discover that the list included preservatives, flavor enhancers, and artificial color.

"There must be something wrong here," I commented. "Remember the national yogurt manufacturer the fruit salesman told us about? Well, I can tell you for a fact that they don't mention preservatives or artificial color or anything like that on *their* label. All they say is 'strawberries,' but they don't say a thing about the chemical additives *in* the strawberries. How can they get away with it?"

Gordon picked up the phone and put my question to the Department of Consumer and Corporate Affairs.

"The strawberry packer was quite correct in mentioning the additives on *his* label," an official told him. "But the yogurt manufacturer is under no obligation to do the same. Strawberries are an additive in yogurt, and therefore must be mentioned as an ingredient. But the artificial color in the strawberries need not appear on the

yogurt label, because if we required all food manufacturers to name the additives in their additives, the lists
of ingredients would stretch from here to China."

We decided to make it clear to the consumers that
our yogurt contained no chemical additives of any kind
by so stating on some brochures that we had Phil pin up
on all the dairy-case shelves in the stores. The next week
he came back from his deliveries looking baffled.

"The store managers told me that the Department
of Consumer and Corporate Affairs confiscated our brochures. Don't ask me why. They apparently went around
to every dairy case and threw them in the garbage."

Gordon picked up the phone again and asked the
Consumer and Corporate Affairs people what they thought
they were doing.

"It's our job to protect the public from fraud and
misinformation," said a sanctimonious voice on the other
end of the line. "You cannot make the claim that your
yogurt contains no chemical additives, since everything
in nature is made up of chemical components. Anything
you eat has chemicals in it."

"Yes, but we're talking about synthetic chemicals,"
Gordon explained. "We don't use artificial flavor, or red
dye number two, and we want the public to know."

"You'd only confuse them. As long as food manufacturers use ingredients approved by the government,
there's no point in trying to claim that one chemical is
superior to another."

Our next run-in with the bureaucrats had to do with
our attempt to put yogurt cheese on the market.

"I'm sorry, but there's no such thing as yogurt cheese,"
said another official in a slightly condescending tone of
voice, as though chiding me for not having read the regulations properly.

"I know. I invented it," I said. "Or at least, I think I did. Anyway, if you say there's no such thing as yogurt cheese, then I must have invented it for Canada at least, and maybe even for North America."

The official was unimpressed. "I'll have to take it under advisement," he sighed. "I've never heard of such a product. There are no standards for it at all. I wouldn't know how to categorize it. Are you sure you want to produce this item?"

"I'm positive. It's a first in the dairy field, don't you see? It'll be hailed as a Canadian initiative, and when it gets popular people will refer to it as Nova Scotia yogurt cheese, and they'll eat it with Nova Scotia smoked salmon. It'll put us on the map!"

"I don't need to be on any maps," he said wearily. "I have enough problems. Anyway, if you insist on going ahead with this, I'll have to look into it. I'll check with my counterparts in the States. Maybe they'll know what to do."

Meanwhile we still had the bills to pay, and our dairy plant was only operating at about 30 percent capacity. If we were not to be allowed to boost our yogurt sales by letting the public know that our product was free of chemical additives, and if our yogurt cheese launch was to be held up indefinitely because of its originality, then our best course of action was to concentrate on selling as much ice cream as possible while the summer lasted. The North Shore Craft Festival, an annual event that took place in mid-June, seemed to be the most obvious place to sell our homemade, hand-churned ice cream.

"No, we can't have you selling your ice cream at the craft festival," said the chairman of the festival's organizing committee. "The purpose of the craft festival is to give local craftsmen an opportunity to sell their crafts.

But we did *not* organize the festival for the purpose of promoting the local business community. So I'm afraid you'll have to peddle your ice cream somewhere else."

"But we're craftsmen too, just like the others," I protested. "Our homemade ice cream is hand-crafted with the same care and skill that any other craftsman uses in creating a product made of wood, or leather, or stained glass."

"I'm sorry, Mrs. Jones, but that kind of reasoning just doesn't wash. If we let businessmen into the craft festival, then pretty soon they'd be selling refrigerators and who knows what else. We have to draw the line somewhere."

"You mean there are hand-crafted refrigerators?" I asked, feeling genuinely puzzled.

"Look, I don't have time for this," she said, giving me a thin smile topped off with an irritated frown.

"Well, what if you sold my ice cream at the barbecue, then? I could sell you as much as you need, and you could mark it up and keep the difference."

"We're going to be selling a commercial brand of ice cream at the barbecue, thank you very much," she said primly.

"But why a commercial brand? Ours is better, and it's homemade. It's more appropriate for a craft festival. You should be promoting food crafts, not Kraft Foods!"

"Mrs. Jones, I'm going to have to be painfully blunt. Your ice cream is too expensive. We can make much more profit with the commercial brand because it costs about half as much as yours."

"But ours weighs twice as much as theirs! Commercial ice cream is full of air, and it has chemicals to keep it puffed up and other chemicals to prevent it from melting, not to mention the artificial color and flavor and the stabilizers and emulsifiers . . ."

As the chairman of the organizing committee of the North Shore Craft Festival showed me to the door, she made it exceedingly clear that she had every intention of ensuring that the visitors continued to be given only commercial ice cream at the festival. My last impression of her was of a woman softened to the melting point as she listened dewy-eyed to a salesman explaining to her how profitable it would be to substitute hot dogs for the chicken that was traditionally served at the barbecue.

As luck would have it, a friend of ours invited us to park our ice-cream truck in his driveway, which happened to be strategically located just across the street from the festival grounds. We put up a big sign indicating that Peninsula Farm ice cream was for sale, and soon the truck was surrounded by eager customers who generously praised the organizers of the craft festival for their perspicacity in making homemade ice cream available to them. We promised to pass the message along.

Our good fortune, however, was short-lived. Soon after we opened for business a representative from the Department of Consumer and Corporate Affairs came along wanting to confiscate our ice-cream sign.

"I'm afraid your product does not meet the standards set by the government for ice cream," he announced rather apologetically.

I was stunned. "But that's impossible!" I exclaimed. "We make our ice cream with nothing but the best ingredients. Fresh, pasteurized cream, farm fresh brown eggs, pure vanilla extract . . ."

"I know, I know," he said, holding up his hand. "But we've had your product tested, and it doesn't contain enough solids to meet the standards. If you want to use the term *ice cream*, then you'll have to change your recipe. You'll have to add extra sugar, or some milk powder, or something along those lines."

"But I don't want to change my recipe," I protested.

"Then you'll have to change the name of your product," he suggested. "You'll have to call it a frozen dessert, or a frozen dairy confection, or whatever you wish, but it doesn't meet the standards set for ice cream in its present form."

"But how could it not be ice cream? It's the only real, homemade ice cream available in Canada today! I'm making it exactly the way they used to make it generations ago when they first invented the name."

"That may be, but the standards today cover the kind of ice cream that is made commercially, so I'm afraid yours just doesn't fit the description. Unless, of course, you change it in some suitable manner."

Gordon saved the day when he drove up to the ice-cream truck bearing a cooler full of yogurt and a sign stating that Peninsula Farm *products* were for sale. I spent the rest of the afternoon asking the visitors if they wanted yogurt or ice cream, always fearful that the official from the Department of Consumer and Corporate Affairs might be lurking somewhere within earshot, ready to pounce on me and haul me off to jail for daring to call my ice cream by its proper name.

Landing Goliath

Ed Shoemaker, the head buyer at Goliath Food Limited, refused to budge an inch on the issue of a 20 percent rebate for the head office. It was a payment that Peninsula Farm could not afford, and yet at the same time we desperately needed to expand our market in order to amortize the cost of our dairy plant and bring it to full production capacity. Every few months I had been paying Mr. Shoemaker a little visit to remind him that our yogurt was enjoying a slow but steady climb in popularity and would be an excellent drawing card for his customers if he would only list it in his chain, but he remained resolutely impervious to my many arguments. I even gave him a petition with the signatures of over six hundred customers who wanted to buy Peninsula Farm yogurt in Goliath stores, but he only gave a grunt of supreme disinterest and had his secretary file it in a back room.

By August I had stockpiled just under two hundred gallons of twelve different flavors of ice cream. I was pinning my hopes on a successful seven-day scooping spree at the South Shore Exhibition, where I had secured some space in the main hall. I felt quite sure I would be able to sell most of my ice cream, for people always flocked to the exhibition from all parts of the province to see the prize farm animals and watch the judging events. My optimism, however, evaporated the moment I arrived to set up my booth. In my usual greenhorn way I had neglected to ask the organizers who my immediate neighbor would be, so my heart sank when I saw a big, professional sign in the booth next to mine announcing the presence of a well-known encyclopedia publisher.

My worst fears were quickly realized. People came pouring into the main hall as soon as the exhibition opened, but when they caught sight of the encyclopedia booth they picked up their pace and scurried away as fast as they could before the encyclopedia salesmen could give them a pitch. A full hour passed before I sold my first ice-cream cone to an intrepid individual who stolidly withstood the pleas of the salesmen in the booth next door. The visitors were definitely not interested in hearing about the eight-hundred-dollar set of encyclopedias that could be purchased for a mere six hundred dollars at the exhibition; nor were they willing to enter their names in a draw for an atlas. They obviously found it much more appealing to buy their ice cream from the booth across the hall, especially since it was flanked by hamburgers and fries on one side and a soft drink stand on the other. I watched sourly as the customers jostled one another impatiently while they waited for my competition to serve them their cones.

"A nickel for your thoughts," came a voice from somewhere behind my left shoulder. "Just trying to keep up with inflation, you know."

I turned to see one of the encyclopedia salesmen smiling at me over the partition.

"I don't think you'd want to know my thoughts just now," I said, trying to muster a sportsmanlike attitude toward the unwelcome turn of events.

"I'm afraid I know them anyway," he replied. "It happens every time. We scare everyone's customers away. What can I say? All I can do is apologize and hope it won't be too bad for you today. It seems a shame, though, that people should take to their heels and run away from knowledge as soon as it rears its ugly head. The truth is, you and I are probably the only ones in the whole place who are selling anything worth buying, and yet the people can't wait to spend their money on junk and garbage. Look at them! Cotton candy, stuffed toys, paper hats, they're wallowing in the stuff!"

It turned out that the encyclopedia man was an executive from the head office in Ontario and was visiting the area in an effort to determine what sales approach might best succeed in persuading Maritimers to buy encyclopedias. I did what I could to provide him with information about Nova Scotia, but I soon realized that his own knowledge of the region was far more complete than mine. We whiled away the time talking about subjects ranging from aardvarks to Zoroastrianism, but I was no match for him. He was clearly a man who knew his product well, but he was also curious to know more about mine, too.

"Your little company intrigues me," he said one

afternoon during the preprandial lull. "Where exactly do you sell your yogurt, anyway?"

"We have it in all the chains," I told him. "Sobeys, Save Easy, No Frills, Capitol, IGA, Dominion, Best for Less, Wade's . . . all of them except Goliath," I added with a touch of bitterness.

"Why doesn't Goliath want to carry it? Are they crazy?"

"Crazy like a fox. They're holding out till I cough up a twenty percent rebate for their head office."

"Twenty percent? You mean over and above the markup between the wholesale and retail price?"

"That's right. And I'll *never* be able to afford a forty percent spread. Not unless I decide to go big, and I don't want to do that. I'd end up compromising my quality, and my customers would never forgive me. I'd also have to get into packaging fluid milk to get my sales up to the point where I could afford to give away forty percent on yogurt, but I don't like that idea either. I'm not interested in packaging milk. I just want to concentrate on yogurt because that's where I can make a real contribution as far as quality is concerned."

"You know who would be interested in this?" said the encyclopedia man when I paused for a moment. "Stuart Ritchie."

I looked at him blankly. "Who's Stuart Ritchie?"

"You've never heard of Stuart Ritchie?" he said, his eyes widening in amazement. "Why, the Ritchie family owns half of Canada. They control the Catchall Corporation, which is an umbrella for hundreds of other companies, both national and international. Goliath Food is just one of them, down on the bottom of the pyramid somewhere."

"But why would Stuart Ritchie be interested in me?" I asked.

"Appearances. Appearances are everything. It doesn't

look good for one of their companies to be treating you like that. After all, you provide jobs down here. If there weren't any jobs, there'd be no customers for their outlets. So the smart companies always cater to local businesses. They know which side of their bread the butter is on."

"Well, I'm sure I'm much too small for Goliath to be concerned about me."

"That's just the point. You're too small for them to be pushing you around, too. Why should they keep you out of their chain just because you can't afford the rebate? Twenty percent of your annual sales is peanuts to them. They shouldn't be quibbling about that kind of money. They're risking their reputation for peanuts. That's what it really boils down to. And that, to my mind, is sheer stupidity."

"No wonder Stuart Ritchie would be interested," I mused. "But how would I ever get past his secretary? If I called him she'd never let me through to him if she didn't know who I was. And she probably never gives him any mail that she doesn't think is important."

"That's no problem. I have his private number. It goes right through to his desk."

"You know Stuart Ritchie?" I said, astonished.

"I know him well enough. We had lunch together just the other day, so I'm probably still fresh in his mind. I run a home for disturbed children, and Ritchie has made several fairly substantial contributions, so we get together about once a year or so and discuss how things are going. That's why I have his number, and if you want me to call him for you I'm sure he'd be happy to talk to you. What do you say?"

"Let's go!"

The encyclopedia man took me to the phone booths located just outside the ox barn, and I proceeded to unload my tales of woe on a mildly bemused Stuart Ritchie while the oxen bellowed vociferously in the background. Ritchie listened politely until I had finished, then spoke to me happily about his memories of the ox pull when he had visited the South Shore Exhibition himself in years past. Just before bidding me good-bye he commiserated with me briefly about what he described as the small misunderstanding that had occurred at Goliath, and he promised he would have it straightened out right away.

The very next morning Ed Shoemaker was on the telephone.

"Good morning. Is this Dr. Jones? Yes, Dr. Jones, it's good to hear your voice again. How is your husband and the family? Good, good. Look, Dr. Jones, I've been thinking . . ."

The following week we made our first delivery to six Goliath stores, with the other forty-three to follow when we felt we were ready. We never did sell much ice cream that summer at the South Shore Exhibition, but later in the year we received a letter from the encyclopedia man telling us that the disturbed children at the home in Toronto were still enjoying the case of "frozen dairy confection" we had sent them.

11

The Cast of Characters Grows

Stars

"Can I talk to you for a moment?" said Phil, leaning through the doorway. I glanced up and saw him standing in the foyer, looking rather out of character in his ill-fitting green suit and carefully parted hair. For a moment I thought he must be on his way to church, but then I remembered it was his day off.

"Sure. Come on in," I said, smiling.

He turned and beckoned to Ginger, and the pair positioned themselves in front of me, their eyes riveted to the floor. They were standing so close to each other that it took me a few moments to realize their hands were locked together in a knuckle-white grip. I should have foreseen it. Propinquity had been working its usual mischief, and my immediate reaction was outrageously selfish. Oh God, what's going to happen to Peninsula Farm if two-thirds of my staff leaves to get married? And just after we landed the Goliath Food account! Couldn't they have chosen a more suitable time to fall in love? Why should they be falling in love anyway, when there's *yogurt* to be made?

"Me and Ginger's getting married next month," Phil blurted out. "We thought we ought to let you know."

I forced my face to light up with apparent joy, while at the same time I spoke sternly to myself. Be good. Feel empathy. Be generous. Remember they've worked hard and they've helped make Peninsula Farm what it is today. And they're young. They have other things on their minds besides yogurt. Let them go gracefully. You can't chain them to the model 500 filling machine.

"I'm happy for you," I said feebly. "And I wish you all the best."

Phil and Ginger were visibly relieved. They were quick to assure me they had every intention of training their replacements before they left, as long as it could all be done before their wedding date. They were planning to build a house of their own on a portion of the parental farm, now that it was suitably refinanced, and they were hoping to be blessed with many sons and heifers. After a short period of adjustment, Gordon and I were able to offer them heartfelt congratulations, but we knew we would deeply miss the two young people who were with us when it all began and who had contributed so much to the growth and development of Peninsula Farm.

During the next few years, we learned a great deal about the fine art of personnel selection and management, but always the hard way, through trial and error. Nothing in our backgrounds had really prepared us for the particular circumstances we were facing in a new country, in a rural area, and in a business with which we were still woefully unfamiliar. We struggled hard to understand the universal principles in our situation so that we could apply as much of our own knowledge and experience as possible in directing the business, but it seemed to us that we continued to muddle along in our usual way, tackling each new problem with a mixture of common sense and bewilderment.

One of our first personnel problems stemmed from the fact that we were eager to hire the very brightest young people we could find. To our astonishment, we discovered that there were many such candidates to choose from, for Peninsula Farm was seen by the younger generation as representing a sort of countercultural operation that opposed the "system" by putting out real food in a world saturated with cheap ersatz substitutes. Hence a good selection of articulate, intelligent young people came to us seeking employment, and we were delighted to welcome them aboard. One of our best workers was a disillusioned dropout from law school, and another exceptional employee turned out to be an English major who had only just learned that the market for her particular specialty was discouragingly thin.

Although many of these overly qualified youngsters turned in a star performance, they never stayed long enough for us to get a proper return on the cost of training them. They would quickly absorb everything we could teach them, but as soon as they came to the end of their learning period, they would inevitably grow restless, and it was never long before they would hit the road again, hoping to find new experiences and fresh challenges over the brow of the next hill.

"The trouble with young people is they have no foresight," Gordon would grumble. "They can't seem to understand anything about the long run. They just live their lives on a day-to-day basis. Why can't they see the potential of Peninsula Farm? Is it me? Have I failed somehow to get the message across? Don't they realize they could build a really interesting future for themselves if they'd only stick it out with me for a while here? They understand what we're trying to do. They all seem to subscribe to the theory that excellent is beautiful, but not one of them will stay to see it through! Whatever happened to good, old-fashioned tenacity?"

Gordon sighed and shook his head. Then he donned his overalls for the three thousandth time and trudged down to the barn to milk the cows.

After a while we became discouraged with the attitude of the bright kids, who basically seemed to be suggesting that it was all right for a couple of older folks like us to do all the uninteresting drudgery, but *they* needed to be kept entertained with constant intellectual stimulation. It was obviously not brains they lacked, but imagination, and there appeared to be little we could do about it.

In order to cut back on the high attrition rate at Peninsula Farm, we decided to concentrate our efforts on hiring personnel who derived more satisfaction from physical exercise than from mental challenges. Michel Vautour was just such an individual. He was a tall, lanky Bunyanesque French-Canadian youth who came from a small town in Quebec renowned for its maple syrup. He was used to working in the woods with the maple trees, but he would like nothing more, he told us, than to learn to run a farm so that he might someday have one of his own. As he was far from home and had nowhere to stay, we agreed to take him in until he found lodgings.

At breakfast the next morning, Valerie and Vicki, who by then had reached the ages of six and four respectively, watched wide-eyed as Michel devoured an entire loaf of bread, half a pound of butter, a pitcher of maple syrup, and half a dozen eggs.

"No wonder his parents sent him down here," whispered Gordon when Michel disappeared into the kitchen to open another quart of milk. "Who could afford to feed him?"

"Now I understand why he wants to get himself a farm," I whispered back.

"How come he's not *fat?*" piped Vicki, gazing after him in wonderment.

"Shut up, Vicki," said Valerie sternly. "He might hear you."

"How come *you* don't eat that much, Dad?" she continued in her high-pitched voice, looking at Gordon with an expression of undisguised disappointment.

"I'd blow up like a balloon," he laughed.

Vicki digested this for a moment. "Then why doesn't Michel blow up?" she asked, her voice reflecting a mixture of awe and tentative alarm.

"He's from *Quebec*, you dummy," said Valerie impatiently.

But if Michel ate enough for three men, he also proved himself to be their equal in his capacity for work. After breakfast Gordon handed him his fence mallet and asked him to pound down all the stakes on the farm. He calculated it would take him at least two days to finish the job, for now that we were harboring a full herd of thirty dairy cows on only twenty-five acres of land, it had been necessary to divide our fields into smaller lots so that the cows would graze more efficiently. The number of stakes to be pounded down had consequently multiplied to the point that fence repair had become the least desirable job on the farm. It was a chore, however, that could not be put off, for to do so would only prolong the number of days the cows remained in the barn busily manufacturing manure in backbreaking quantities.

After a lunch of six hamburgers, a family-size bowl of salad, a quart of milk, three portions of maple yogurt, and an entire cherry pie, Michel turned to Gordon and asked him what he wanted him to do for the afternoon.

"Just keep on pounding fence stakes," said Gordon, relieved to be rid of the responsibility.

"The fence stakes, they are all down good," said Michel proudly.

Gordon stared at him. "*All* the fence stakes? You can't have pounded down over a thousand stakes. Not in just one morning!"

"Come and see," said Michel, grinning from ear to ear.

Every last one of the stakes was firmly in the ground, and even the wires had been tightened and refastened. By early evening the water fences had been made, the pigpen had been cleaned, and one end of the barn had received a first coat of scarlet paint.

"Now we milk the cows," said Michel, striding eagerly into the barn with Gordon in tow.

Under normal circumstances, two people in a dairy barn can proceed at an even pace if one of them milks the cows while the other shovels out the stalls, replaces the bedding, and fills the mangers. But Michel's strength and his idea of a comfortable tempo came nowhere close to the realm of normality. He snatched up a shovel and briskly scraped the dung into the troughs behind each line of cows; then he simply dug the shovel into one end of the manure and walked effortlessly down the entire length of the trough until he and half the manure were on the other end of the barn. When the time came to feed the cows, Michel didn't bother to use gentle pushes and friendly persuasion to separate them so he could reach their mangers. Instead, he marched up to the end cow and sent the whole line flying with a simple hip check. And as for changing their bedding, he saw no point in promenading around each cow and forcing his way between them again. His legs were so long and stiltlike that all he had to do was walk right over their backs, while the startled animals stared over their shoulders at the human beanpole playing leapfrog across their spines. Gordon had only milked four cows when Michel announced he was finished with his half of the work and was going to paint the south end of the barn.

Years later, after he had left us to go on to bigger and better things, Gordon received a visit from an army officer. Apparently Michel had applied to join the service, and the officer wanted to know if Gordon thought he would make a good recruit. Gordon didn't hesitate for an instant. "Sign him up right away," he advised the officer. "Michel Vautour is one man you wouldn't want to have fighting on the other side."

Extras

As time went by, we began to realize that bovine management and personnel management had certain points in common. When a cow could not produce enough milk to pay for her feed, we had no choice but to cull her out of the herd. By the same token, we tried to make it crystal-clear to our employees that our business would not survive unless we all pulled together and did everything possible to produce not necessarily the most yogurt, but definitely the best yogurt on the market.

It was not always easy to get this message across. We started out offering carrots by way of frequent raises in pay for outstanding performance, but we soon learned that many members of the staff had a disconcerting tendency to accommodate themselves quickly to their new salary levels and to take it for granted that no further improvement was necessary on their part.

Next we tried to inspire them with bonuses for productivity and quality control, but this only created outbreaks of heated arguments over who should be getting what for how much. We were also dismayed to discover that the hourly wage earners were doing all they could to drag out their work for as long as possible. The length of the workday was miraculously reduced by as many as three to four hours when we switched them over to a straight salary, but to our horror we found out that they were using thoroughly unacceptable shortcuts that were causing noticeable differences in the quality of the yogurt.

I eventually abandoned the carrot and resorted to the stick method of personnel management, but this only made them sullen and craftier than ever at figuring out ways of getting around old Simon Legree. It was then that Gordon and I finally understood that the bad apples simply had to be culled from the staff without delay, but it was never as easy for us to fire an employee as it was to take a cow to auction.

Gordon would sometimes grow despairing. "I feel like Henry Higgins in *My Fair Lady,*" he would say. "Remember that song he sang? The one where he kept asking why a woman couldn't be more like a man? Well I feel the same way, only I keep wondering why my employees can't be more like *me*. Why don't they give me their best effort? Why don't they take pride in their work? Why can't they share my vision of a company determined to achieve excellence?"

"Maybe because they don't own the company," I suggested.

Gordon groaned. "Great. Just great. They'd like to have a piece of the action, but where were they when I needed them? Where were they when I was milking cows at midnight in a blizzard with the flu and no electricity? And where are they now when I have to pay the bills? And what about risk? Would they be willing to risk everything they own to get a business on its feet? Hah! They want to own the company? I'll tell them what it's like to own a company."

"I guess it's just human nature to want to get paid as much as possible for doing as little as possible."

Gordon promptly exploded. "If that's human nature, then take it away. I don't want any human nature at Peninsula Farm. I want heroes. Show me an employee who puts the company first, and I'll show you a boss who puts the employee first. But none of this flash-in-the-pan stuff. I don't want a bunch of bees buzzing around here and then leaving to go to other flowers. And I don't want a whole pile of queen bees lounging around either, thinking that the world owes them an easy living. You know who our biggest competitor is? Unemployment insurance. No wonder we have such trouble finding anyone good!"

The plot thickens

Gordon was fond of reminding me that ten times nothing is nothing, and I had to agree with him that his equation applied all too often to the staff at Peninsula Farm in those early years. It sometimes appeared to us that we had more than our fair share of misfits, and they often came disguised as something else. One young man quickly caught on to Gordon's desire to find a loyal, dedicated lieutenant, so he worked hard to create the impression of being a faithful employee with the company's best interests at heart. We were both duly impressed. In fact, our hopes for this fine, clean-cut young man had risen so high that we were seriously contemplating grooming him for a management position until we unexpectedly discovered that while he was busy polishing our apples he was also losing no opportunity to rob us blind.

Others in our dramatis personae seemed to have flaws that condemned them to failure. Our most talented driver was superb with the trucks, accurate with his figures, and had the gift of cheering up even the grumpiest receivers at the stores. Everything went well until we learned, to our immense disappointment, that he was an alcoholic. We did our best to persuade him to accept treatment, but he was convinced that his drinking was under control, and there was nothing we could do to change his mind.

His replacement was also a high-spirited fellow who was competent at his work and well-liked in the stores, but he was such a speed demon that it wasn't long before the Mounties caught up with him and relieved him of his license.

In my effort not to get involved with another speeder, I went a little too far the other way and made the mistake of hiring a jobless anthropology major who had spent most of his adult years in the library. After four days of delivering yogurt, he quit suddenly, claiming to be on the verge of collapse. He would not stay until I found a new driver, he said, for he was leaving the next day for Europe, where he was looking forward to enjoying a rest cure at a vegetarian resort somewhere in the south of France. I vowed the next driver I hired would be an avid steak-and-potatoes man.

That steak-and-potatoes man turned out to be none other than myself, for whenever disaster struck at Peninsula Farm either Gordon or I had to jump into the driver's seat and do the necessary. I was the only available truck driver at the time, for Michel had already gone on to greener pastures and Gordon had to milk the cows, so I started off early the next morning taking Valerie with me for company. But Valerie, who was just entering her eighth summer, was not interested in providing only company. She was determined to work right along next to me, and she begged me to train her to take on part of the job. I reluctantly agreed to let her "help" me by picking the yogurt off the shelves in the truck while I went into the stores and tracked down the receivers.

To my great surprise the plan worked perfectly. By the time I found the receivers (who were forever on their coffee breaks) and dragged them to the back door, Valerie had already piled the correct amount of yogurt onto a pushcart and was waiting patiently to be let in. If the receivers and I were delayed long enough in the stores, she would even have most of the yogurt priced as well. Once we were inside again she would push the cart to the dairy case and we would put the yogurt on the shelves, making sure all the expiry dates were properly rotated. Then we would march out with our mountain of empty boxes and negotiate our way through the traffic to our next stop.

We would usually be well ahead of schedule by lunchtime, and by midafternoon the day's deliveries would be over. Our teamwork proved to be a great success, and I told Valerie how enormously proud I was of her for making my job so much easier. She had easily outshone the previous three drivers because she had taken a real interest in her work and had been determined to do well.

We were chatting happily together that afternoon as we drove through the crowded streets of Halifax when suddenly Valerie saw something that excited her interest.

"Oh *look*, Mum!" she exclaimed. "Look, a *playground!* Do you think we could stop for a while so I can play on the swings?"

The day had gone so smoothly and the yogurt had been delivered so accurately that I had completely forgotten my partner was only seven years old.

Auditions

Painful as it was, a rigorous culling process proved to be invaluable to us in putting together a winning team, but there was often a heavy price to pay. Every time we fired someone the whole company seemed to teeter on the verge of calamity until at last the replacement was fully trained. It was as if an actor suddenly disappeared just before a dress rehearsal or even a performance, leaving the rest of the cast members to ad-lib their way through the show. The worst of it was that there were no understudies standing in the wings waiting to take the place of the absentee, so Gordon or I would quickly have to don an appropriate costume and take over the role. This meant that the play was temporarily without a full-time director, which had its advantages and disadvantages. All sorts of difficulties arose when the group moved along without orchestration. The telephone could not be answered on time, orders were misplaced, trucks were loaded incorrectly, we ran out of ingredients, and health officials would inevitably show up to do plant inspections right in the middle of the general chaos. On the other hand, Gordon and I benefited from being on stage with the others, for the hands-on experience kept us up-to-date with all aspects of the business. Our employees benefited too. We were almost always deeply impressed with the performances of our veteran actors, and after working cheek-by-jowl with them all day in the factory we would crawl back to the house, panting and exhausted, and give them all a raise.

It finally dawned on us that we could save ourselves a lot of trouble if we concentrated harder on hiring the right people instead of having to suffer the consequences of firing the wrong ones. We began to subject all candidates to a rigorous audition, which began from the moment they applied for a job. Anyone who telephoned and said "Ya hirin' on?" was automatically disqualified, as were those whose mommies called to inquire about the job opening. When it came to interpreting the significance of spelling errors on application forms, Gordon and I had some differences of opinion. At first Gordon expressed the view that good spelling had nothing to do with making good yogurt, but I felt strongly that a sense of accuracy was needed in measuring ingredients and calculating the ideal time and temperature of incubation according to the strength of the starter. We finally agreed that candidates unable to spell the names of their streets or schools would probably be better off doing something else.

We eventually decided that attitude was unequivocally the most important quality of a successful applicant. Corny as it may sound, we were always impressed with a candidate who was more interested in what he could do for the company than what the company could do for him. Now that I was sitting on the other side of the desk, I winced at the thought of my own approach to job interviews back in the days when I pounded the pavements in New York. I had tended to emphasize how fortunate my potential employer would be to land a candidate like my unmatchable self, and I took it for granted that hard bargaining on questions of salary and dark hints of better offers elsewhere would make me seem ultradesirable. Now when applicants told me they could make more money or work fewer hours or get better benefits at some other company, I wished them luck and sent them on their way. Not that we wanted to be stingy or hold back on our employees. One of our greatest ambitions was to reach the point where we could compete with any company in the province on salaries, benefits, and working conditions, but we fully expected the employees themselves to help us achieve that goal, and in so doing, prove themselves worthy of sharing the rewards.

One of the most important reasons for auditioning the new candidates so carefully was to protect our regular staff from too much exposure to people with the wrong attitude, which tended to have a demoralizing effect on everyone. One new fellow bombed into the factory on his first day of work and turned up the radio full blast, announcing that everybody should loosen up and be cool. He bebopped to rock music all day while everyone else had to run around after him in a desperate attempt to stop him from manhandling warm yogurt, or dropping three-hundred-dollar thermometers, or endangering every delicate instrument or piece of machinery that happened to get in his way.

Another new worker turned out to be a rather spoiled young woman who evidently thought she was too good to wash pots or load yogurt onto trucks. She spent most of her first and only day sitting on a counter watching other people work, never imagining that she should lend a hand unless specifically directed to do so.

Still another new recruit managed to make so many mistakes on her first day that production was held up by about two hours, but when five o'clock came she got her coat and drove away, leaving the others to finish up on their own.

Needless to say, it was hard on the rest of the employees to be told, directly or indirectly, that making yogurt was uncool, or that washing pots was demeaning, or that smart people know their rights and never hesitate to escape at the sound of a bell. And yet they managed to shrug it off with good humor, quick to point out that they, the survivors of the unpleasant but necessary culling process, were still working at Peninsula Farm instead of jiving to a ghetto blaster on an unemployment line somewhere. Gordon and I were proud of our hard core of tried-and-true staff, and we did our best to praise them as much as we could, knowing that without such encouragement their healthy attitude toward their work could not be expected to last.

Nothing brought this home to me so clearly as when the shoe was on the other foot. One day I went dancing into the office of my supervisor, delightedly holding aloft my *second* published book for his inspection. I had spent the better part of three years researching the project, and had even been honored with a Canada Council grant to defray some of the costs, but it only took my supervisor two minutes to glance at it and lay it aside with a disgruntled sniff.

"That's very nice, Mrs. Jones, but the book you wrote is part of a *series*. I've read some of the other books in that series, and I must say I wasn't overly impressed."

"But how can the series itself contaminate an individual book?" I asked, straining to understand his logic. "Surely each book must be as good as its author!"

But he was apparently either unwilling or unable to judge my book on its own merits. "It's simply not scholarly enough," he complained, flipping the pages in a distracted way. "The language you use, the style of your writing . . . Well, you could hardly call it *academic* by any stretch of the imagination."

I felt thoroughly deflated and more than a little angry as I hurried down the hall to teach my next class. Even though I kept reminding myself that I could get along very nicely without his approval, and despite my attempts to convince myself that I didn't need to be looking for encouragement from professorial types, I still couldn't help feeling that it would have made a big difference to me if he had managed to twist his thin lips into a mildly encouraging smile, or even squeeze a small sparkle of appreciation from his narrowly critical eye.

I had learned my lesson. Everyone needs a pat on the back once in a while, and certainly we all enjoy being given the recognition we deserve. I decided then and there to throw a big party for the staff at Peninsula Farm, to celebrate the successful efforts of the little team that made the world's best yogurt.

12

Distribution

Yogurt wars

As our yogurt sales increased and our market share expanded, we began to detect signs of irritation on the part of the competition. The mouse was daring to roar at the lion, and the lion was distinctly unamused. It wasn't long before one of the dairies decided to drop its prices, and the others immediately followed suit, thinking it would only be a matter of time before the annoying little rodent quietly crawled back to its hole. At the first sign of the price war, I ran home to Gordon and started squeaking in terror.

"Gordon, we'll *never* be able to match their prices! They're selling their yogurt at less than our cost! How can they *do* that?"

"Well, for one thing they can afford to lose money on yogurt because they have fluid milk and several dozen other dairy products to push while the war is on. And for another thing, many dairies buy their yogurt from Quebec and Ontario, and I'm sure it's dirt cheap. No wonder they can sell it for next to nothing if they feel like it."

"But Gordon, those other dairies are going to put us out of business, and you're just standing there, cool as a cucumber, explaining to me all about how they have the edge and will *always* have the edge! How can you be so calm?"

"It won't do us much good to get upset," he laughed. "And anyway, that's just what they want. We'd be playing right into their hands. They'd *love* to see us fold up our tents and creep away, but what they don't know is that we're here to stay. And we're going to win, too."

"How?" I croaked.

"Just by hanging in there, that's all. We're not going to panic. We've always said we make the best yogurt in the world, so now is the time to put it to the test. If enough people agree with us, they'll buy it anyway."

For the next few weeks, I lived with my heart in my throat. At least, it would make a leap for my throat whenever I turned my attention to our sales figures, but fortunately my textbook publisher was pressuring me to write a second edition as quickly as possible, so I was happy to distract myself with that task while the yogurt war continued to rage in the background. Every once in a while I would take a break from writing (or preparing classes, or correcting papers, or chauffeuring the girls to their various destinations) to sneak a timorous peak at our bookkeeper's neatly printed ledger, only to discover that our sales were not dipping anything like as much as I had feared.

"You see?" said Gordon. "All that worrying for nothing. You should save your energy."

My anxiety had turned to immense relief, and the relief was in turn yielding to a strange rush of what could only be described as love for the Unknown Customer who had loyally continued to buy our yogurt in spite of the assorted temptations provided by our competitors in the form of giveaways, coupons, two-for-one sales, and free samples. As a final irony, Gordon and I were both surprised to notice that our sales went up considerably when the competition finally decided to pull in their horns and call off the price war. The other dairies had evidently attracted a lot of attention to yogurt in general during the flurry of activity they had generated, and we found we had *all* gained market share once the dust had settled. We ended up better off than ever before, and we gratefully took our hats off to our competitors for stirring up so much excitement at a cost we never could have afforded.

But the beast was not about to go back to its lair. If the dairies could not tempt our customers away from us, then they would see to it they had a hard time finding our yogurt in the stores. Their first approach was to make sure their yogurt was on all the eye-level shelves in the dairy cases everywhere. Even I was confused when I trotted in for store-checks and found our yogurt missing. At first I thought our sales had improved so much that the stores had completely sold out, but on further scrutiny it eventually dawned on me that our containers had been moved to the bottom shelf of the dairy case.

"What's going on?" I asked a grocery clerk, pointing to our low-lying yogurt.

He shrugged. "We get our instructions from head office. It looks like those other guys have found a way to sweeten the deal. But don't worry, your customers will find it down there sooner or later. They'll have to. They'll probably *trip* on it."

The grocery clerk was right. After a brief dip in sales, the customers finally discovered our yogurt in its hiding place, and soon business was back to normal again. But that was only the beginning. The next thing I knew there were mountains of butter, cottage cheese, and other assorted dairy products piled solidly right in front of my yogurt so that it couldn't be seen from any angle at all. The up-front yogurt price war had turned into a series of sneaky guerrilla skirmishes that were very hard to control.

"I'll do my best," said a dairy-case boy when I brought

the situation to his attention. "But I can't promise anything. The salesmen are on commission, you know, and they'll do anything to increase their sales. I know it's dirty pool, but I can't stand around watching all the time. I'm overworked as it is."

"You're kind of at a disadvantage," said another dairy-case boy in a store around the corner. "Those guys come in here every day to stock their milk, so it's no trick for them to mess around with your yogurt. I'll tell them to lay off, but I'm busy with my own stuff. You ought to come in every day too, if you really want to defend your shelf space."

But I knew that short of cloning myself or hiring an army of merchandisers, it would be impossible to be everywhere at once. The situation was particularly difficult in some of the smaller stores, where my yogurt was usually kept in a glass-doored cooler along with milk and other dairy products. I began to notice that Peninsula Farm yogurt was gradually disappearing from the coolers and turning up unexpectedly among the dips, the imported cheese, and even in the vegetable displays. When I expressed my dismay to the grocery managers, they told me that the coolers were owned by the dairies and that they had lost no time in exercising their proprietary rights by instructing the stores to remove my product from their coolers without delay. It seemed unfair to me that the large dairies were able to control my sales by purchasing the refrigeration equipment, but the store owners were all vehemently in favor of the idea as it saved them from having to buy coolers of their own. I went home that night feeling depressed and angry.

"You're never going to get them to call off their dogs," said Gordon, still relatively unperturbed by this latest development. "So if you can't beat them, what do you say we join them?"

"*Join* them? How?" I asked, horrified. "Let them buy us out? Let them make our yogurt under contract?"

"No, no," he said, smiling and shaking his head. "Nothing like that. I was thinking of distribution, that's all. If those fellows are in the stores every day, and if they're so hell-bent on selling yogurt, then let them sell *ours*. After all, we're buying half our milk from them as it is. And now that we've outgrown our thirty cows, the dairies will be selling us more and more milk as time goes by. So they really ought to be helping us rather than trying to grind us into the dust. That's the way *I* see it, anyway. They'd be doing their farmers a favor by dropping that dopey Quebec yogurt and concentrating on ours for a change! Because why should the local dairies be supporting the farmers in Quebec when their own cows need jobs?"

Gordon's proposal made perfect sense to me. The other dairies sent their trucks all over the province, and serviced far more stores than we did. If we were to join forces with them, we could expand our market into new territories and continue to amortize the constantly growing costs of new equipment, extra trucks, additional staff, and ongoing repairs. It was easy to see from our ledgers that distribution represented a cost to us that was out of proportion with the rest. Because we manufactured only one product, our average delivery came to a mere forty or fifty dollars, whereas other dairies were dropping hundreds and sometimes thousands of dollars worth of milk and by-products at the stores. Clearly their delivery costs could not be nearly as high as ours, and since only one other dairy in the Atlantic provinces produced yogurt at all, we felt there would be no conflict of interest if we suggested to the others that they distribute our yogurt along with their own dairy products.

"I can see what you're getting at," said the president of one of the major dairies, "but we're better off staying with the commercial yogurt we bring in from Quebec. People are used to their brand name down here, so our volume is higher with them than it would be with you. Don't get me wrong, though. Your yogurt is much better than the one we distribute. My wife eats it all the time. But let's face it. We've got to stick with the high-volume item."

The general manager of another large dairy had similar views. "There's no two ways about it. It's more profitable for us to distribute the Quebec yogurt than it would be to deliver yours. Do you know how many farmers ship their milk to that plant? Eight thousand! How's *that* for size? They've just about wrapped up the whole province of Quebec. They can afford to sell it to us for next to nothing because they've got such a big market at home. I know your yogurt is better, but take my word for it, profit is the name of the game. We need the extra margin we get on the Quebec yogurt because that way we can move some of it in the right direction, you know what I mean? Money talks, but it sure doesn't grow on trees. We have to take it where we can find it."

The owner of still another large dairy had a slightly different point of view. He shrewdly agreed to distribute our yogurt on condition that we allow him to continue delivering the Quebec yogurt he already handled. That way if our sales went up at the expense of the Quebec manufacturer, at least he would get something from both ends. We decided to walk away from that deal, however, as we couldn't rid our suspicious minds of the thought that the dairy in question might just possibly be tempted to *make* our sales go down in order to benefit from the higher margin on the Quebec product.

Wherever we turned, we found that the supremacy of the bottom line was ensuring that the market remain forever saturated with cheap goods whose inferior quality was ceasing to surprise anybody. What *did* surprise me, however, was the alarming number of shoppers who seemed not to care. Many of them were apparently content to take home a so-called bargain no matter how poor the quality of the merchandise. I realized, of course, that there existed a hard core of discerning customers who understood the difference; otherwise we would never have survived the yogurt wars. But as we drove home from our meetings with the dairy managers that day, I could not help but wonder just how large this segment of enlightened citizens might be, and whether indeed it would be sizable enough to support our company in its necessary growth.

Losing control

Another hard look at our ledger quickly revealed that our sales had been discouragingly flat for a number of months. We were well aware that we would have to move more product in order to meet our ever increasing costs, but just how to accomplish this was difficult to determine. We had no money to spend on advertising and promotions; nor could we easily afford more trucks and drivers to open up new territory. Just as I was beginning to think we had reached an impasse, we received a telephone call from the general manager of the Sycamore Dairy, which was located in a major city geographically beyond our reach. He had heard through the grapevine that we were looking for a distributor, and he proposed that we allow his company to deliver our yogurt throughout the city of Mirage. He went on to explain that no fewer than four separate dairies in his area were distributing the same brand of Quebec yogurt to the stores, and the situation was getting out of control. They had all agreed to divide up the stores between them, but since every driver was eager to sell as much yogurt as possible, he would often drop some at a competitor's store while he delivered his milk. This caused total confusion. The four dairies were engaged in constant arguments about who delivered what yogurt where, and nobody could ever agree on who was responsible for returns. The obvious solution for the Sycamore Dairy was to distribute a different brand of yogurt. The owner would have preferred, of course, to represent one of the Quebec dairies, but as they were already being widely distributed by other companies in Mirage, he had been willing to settle for ours.

It seemed like an ideal situation for Peninsula Farm. The Sycamore Dairy had a truck in Halifax every week, and it was to their advantage to pick up their supply from us there so as to avoid *dead-hauling* (returning with an empty truck). They would charge us their usual distribution fee for delivering our yogurt to the stores in Mirage, and our only responsibility was to provide them with the cargo on time and in good condition. They would be responsible for ordering the product, making the deliveries, stocking the shelves, rotating the dates, and crediting the stores for expired product. We all shook hands on the deal, feeling confident that both we and Sycamore would benefit greatly from these mutually satisfactory arrangements.

Sycamore's first order was large enough to provide our staff with an entire day's work. Everyone was in a celebratory mood. Our employees were delighted to have an extra day's pay, Gordon and I were happy to be producing so much yogurt at such a reasonable cost, and our milk supplier was pleased to unload his tanker into our holding vats instead of having to make the long trip back to his own dairy.

The next week, however, Sycamore ordered only one quarter of the original amount, and in puzzling flavor ratios. Instead of ordering extra quantities of such popular varieties as strawberry, peach, and raspberry, they wanted a full complement of flavors with particular stress on maple, pear, and pineapple. Their propensity for making radical changes in the size and ratio of their orders kept us on a seesaw for the next few weeks. We never knew if there would be an extra day's work or not until forty-eight hours before the fact, but worst of all we ran into serious difficulties in ordering ingredients and having them on hand in time for production. Everyone began to feel the strain.

Gordon and I decided it was time to take a little trip to Mirage for a store-check. No sooner did we identify ourselves to the managers than they voiced their complaints in no uncertain terms.

"Look over here," said one irate grocery manager, leading us firmly to the dairy case. "What do you call *that?*"

The store had been kind enough to allocate us a four-foot shelf, which Sycamore saw fit to stock entirely with one flavor—strawberry. Our whole section of the dairy case was nothing but one splash of red containers, whereas all the other sections were well stocked with a full complement of flavors brought in by the competition.

The grocery manager in the next store turned purple when he saw us. "This is the third week in a row that Sycamore has forgotten to leave yogurt. How do they expect us to sell it if they don't give us any? The driver says he doesn't have any left by the time he gets to my store. But look at this. See what I mean? Your section of the dairy case is completely empty. We don't want to pay to refrigerate *air!*"

We were greeted by an equally exasperated manager in still another store. "Sycamore is unloading more yogurt than we can handle here! Most of it is in the walk-in cooler taking up space, and we have to move it around every time we go in for stock. There's mountains of it back there, and it's always in the way."

After apologizing profusely to all the store personnel, we decided to spend the rest of the afternoon with the distribution manager at Sycamore. We carefully explained to him how to use our inventory-control system while he listened with polite disinterest. When we had finished, he leaned back in his chair and folded his arms.

"First of all," he said, lighting a cigarette and snapping his lighter closed, "you've got to realize that our truck drivers would never take the time to write down all those numbers and do all that figuring. They're *truck* drivers, anyway, not mathematicians. And second of all, they've got milk to deliver, and a lot of the time that milk is stacked right in front of the by-products, so they can't always pull out the flavors they're looking for. Then third of all, there's no way for the loaders to know what the drivers are going to need for yogurt. The merchandisers are supposed to keep an eye on the orders and help them out, but they don't have crystal balls. They can only do what they can. I'll talk to them, though, if it'll make you happy."

The next thing we knew warm weather was upon us, and with it came a series of frantic phone calls from the Sycamore Dairy.

"Your yogurt's no good," said the distribution manager, barely able to disguise the note of triumph in his voice. "The stuff is spoiling all over the place. You've got a problem, and you better find it quick."

"How could it be our problem?" I objected. "Our yogurt isn't spoiling in any of the stores *we* go to!"

"Must be something wrong with the refrigeration unit on your truck," he persisted. "Or maybe your plant is unsanitary. You'd better check it out. All I know is you said you'd give us the yogurt in good condition, but it's not in the stores three days before it spoils. It's supposed to last till the 'best before' date, you know. We can't be paying for yogurt that spoils before the code."

Even though it was completely illogical to assume the cause of the spoilage must lie with us when the yogurt we delivered to our own stores was in perfect condition up until and sometimes even beyond the expiry date, Gordon and I felt it would be prudent to check out everything on our end before making any accusations. We brought in the health department to go over the plant with a fine-tooth comb, and we came out with a clean bill of health. We also summoned a mechanic to examine our refrigeration units. He assured us they were in perfect working order and were keeping the truck boxes at a steady 35 to 40 degrees Fahrenheit. There was nothing for it but to take another trip to Mirage to see for ourselves how our yogurt was being handled.

What we discovered put us in a state of shock. The Sycamore trucks were backing up to the loading platforms and holding their doors open for as long as it happened to take the drivers to unload the milk and by-products. The deliveries could take well over an hour if the drop was particularly large or if the receiver was on a break or otherwise occupied. Meanwhile the truck was sitting in the broiling sun while the refrigeration unit labored away in vain, trying to cool the outside world through the open doors at the back. When the driver finally unloaded our yogurt, it was often left on a pallet while driver and receiver discussed the results of the latest ball game. On one occasion, we discovered the yogurt still on a pallet outside the cooler, waiting until such time as the dairy-case boy returned from lunch. When the pallet was finally wheeled into the store, it would remain at room temperature until the grocery clerk found time to put it on the shelves. By the end of the day, Gordon and I had calculated that our yogurt was spending most of the morning in an essentially unrefrigerated truck, up to one hour in the sun on the loading docks, and several more hours at room temperature before it was stocked on the shelves.

"So *you're* the people from Peninsula Farm, are you?" said an angry grocery manager when we dared to introduce ourselves. "Well, you might as well not bother to send us any more of that stuff. I don't know what you're doing to it, but last week all your yogurt *exploded* right there on the shelf! The dairy-case boy had some mess to clean up, let me tell you!"

He stood there glaring at us as we did our best to explain to him how it was that fresh fruit ferments when exposed to excessive heat for long periods of time, but he wasn't interested in hearing the details of why our yogurt was trying to turn itself into champagne. All he wanted from us was the assurance that the product would never again appear in his store.

Within the next two days, we had received urgent phone calls from the managers of fifteen or twenty stores, all with the same complaint. Apparently our yogurt was exploding all over the otherwise peaceful city of Mirage. We knew right away that the game was over. We quickly severed our relationship with the Sycamore Dairy and left the territory feeling mortified. We had suffered an ignominious defeat, and we could only hope that the hapless customers would soon forget their harrowing experience with Peninsula Farm.

Harnessing new horsepower

"That's the worst of it," said Gordon, once we had had time to come to terms with the Mirage fiasco. "It's *our* name on the package, and nobody else's. When something goes wrong with the product, *we* get the blame, no matter whose fault it really is. Nobody else is ever going to handle it with the same care and concern as we do. We may as well get that straight right now."

"What really bothered me," I muttered, "was that Sycamore kept saying our yogurt was the only one that had ever given them any trouble; therefore there *had* to be something wrong with it or it wouldn't have behaved that way."

"There was something *right* with it; that's why it gave them trouble. It's like anything else. When bread has been so chemicalized that it can't go moldy anymore, you can be sure they've got it sanitized to the point where it's not worth eating, either. And yogurt is the same thing. They cook the life out of the fruit they use so it can't ferment, but what do they end up with? Anyone who can't tell the difference has no taste buds at all."

"I can see why they have to do it, though," I conceded. "Now that I know how refrigerated goods are handled by the distributors and by the stores, I suppose the big companies have no choice but to load their food with preservatives and everything else under the sun so they can extend the shelf life as much as possible."

"If that's the case," said Gordon, "we're going to have to take care of our yogurt ourselves, every single step of the way, or we're dead. At first I thought we had to do it that way because we were too small to interest a distributor, but now I can safely say that no distributor will ever be good enough to interest *me!*"

"Making yogurt with fresh milk and real fruit has other advantages too," I added. "If it's true that there's nobody out there capable of handling it properly, then nobody's going to be able to imitate us, either."

"Maybe they wouldn't want to imitate us anyway," said Gordon drily. "I have a strange feeling that if we get out of this alive we may wish we'd been in the business of manufacturing a product that had no shelf life and needed no refrigeration. Boots, or bowling balls, or something solid like that."

As time went by and we continued to grow, we slowly acquired a small fleet of trucks, but no matter how much we struggled to improve our efficiency in the trucking area, distribution still remained our greatest single expense. Diesel trucks cost a small fortune when they broke down, and they could sometimes be the very devil to start on a cold morning. It didn't seem to matter whether we leased them, rented them, or bought them outright; they still seemed to take a fiendish delight in draining our resources at every turn. Neither Gordon nor I understood anything about trucks, and we resented having to direct our time and attention to questions relating to solenoid switches and universal joints. Gordon was by nature an entrepreneur and a superb salesman, and my interests lay primarily in teaching and tasting, but as the business expanded we were constantly required to give up what we liked to do and concentrate instead on subjects for which we really had very little talent and absolutely no expertise.

What surprised us most of all was that a sizable portion of our distribution costs was generated by the stores themselves, but as far as we could determine these expenses were totally unnecessary. All too often a driver would be held up by a breakdown of some sort, only to be told by the store clerks that he had arrived five minutes too late for receiving. We would either have to send an enormous tractor-trailer back to the store in question the next day, or we would have to lose the potential sales for that week. At other times the dairy-case boy would solemnly promise to stock the store shelf with the excess yogurt in the back cooler as soon as there was room for it, but the driver would often return the next week to find the shelf empty and the yogurt still stacked in the walk-in cooler. It appeared that the personnel in the stores could not be expected to do five minutes' work, even though the resulting loss in sales was immeasurable.

The situation was even worse, however, in the unionized stores, where our drivers were not allowed to stock the shelves at all. It seemed as though the clerks were vying with one another to see who could do the least work, for to do otherwise would mean that the rest of the employees would be expected to improve their performance. Our worst loss in sales was incurred in one union store where the employees were trying to prove to the management that they were so overworked they couldn't put the yogurt up at all. The shelves remained totally empty for three weeks until the manager finally agreed to hire two more clerks. My driver came home one day to tell me that our yogurt was selling well now that it was actually being made available to the customers. We, of course, were responsible for all the yogurt that had expired in the back cooler when the clerks had refused to put it on the shelf. The clerks were happy now, he added, for they were able to play poker in the back room with all their spare time. The fact that they had bought their extra time at the expense of Peninsula Farm was not lost on either me or my disgusted driver.

"You can bet you'll never find the *Japanese* playing poker on company time," said Gordon, when we told him what had happened. "This whole country will go down the tubes if people keep trying to do their worst instead of their best. Why don't people automatically take it for granted that they should always do their best, at all times. *Period!*"

Gordon's observations reminded me of how my father would complain that the world was a vast desert of rampant mediocrity. When I was a child I used to think he was exaggerating, but now that I had a few years under my belt I was beginning to suspect that maybe his apoplectic comments were really understatements after all.

13

Caught on the Escalator

Yogurt cheese makes its debut

Gordon and I both knew that if we were to be our own distributors we would have to do everything possible to increase the size of our drops in order to bring down the cost of deliveries. It seemed like the perfect moment to launch a new product. Christmas would soon be upon us, and I could easily visualize our yogurt cheese being served in silver bowls surrounded by chips, crackers, and fresh vegetables, all presented enticingly on a coffee table before a crackling fire. It was time to give the Department of Agriculture and the Department of Consumer and Corporate Affairs my full attention.

After several weeks of letters and telephone calls, the long-awaited approval finally came through. The authorities had thoroughly researched the situation and found that no other company was marketing such a product in Canada, and as far as they could determine, their counterparts in the United States had no knowledge of it either. They had also discovered, however, that yogurt cheese was by no means a product of my own invention, for concentrated yogurt (known as *labneh*) had for centuries been a popular food in the Balkans and in the Middle East.

160

"It looks as though I've reinvented the wheel again," I sighed.

But there was no time to worry about the origins of yogurt cheese, although I was fascinated by some of the details of its history and development. Apparently nomads had stumbled on the idea of concentrating yogurt when they learned to carry the fermented milk in animal skins. The whey then slowly leaked through these porous bags as they were transported across the desert, and by the time the camel drivers unwrapped them at the next oasis the yogurt had transformed itself into balls of cheese. Not only did this new product add welcome variety to the diet of the desert dwellers, but it provided the extra benefit of lasting much longer in the saddlebag because of the higher concentration of lactic acid. Down through the ages other generations of Middle Easterners developed new ways of preserving it even longer by smoking it, drying it in the sun, and covering it with olive oil or tallow. While I doubted that such exotic procedures would go over very well in the Maritimes, Gordon and I were delighted finally to have been granted clearance to produce our yogurt cheese in time for the Christmas season.

As we experimented with the product in the weeks just before the scheduled launch, we reconfirmed the nomads' discovery that yogurt cheese has a shelf life at least twice as long as that of ordinary plain yogurt. We were also pleased to learn that the addition of herbs and salt extended its life span by another few days. We lost the extra days, however, when we decided to tone down the cheese by adding some pasteurized cream, for we were convinced that the public would accept it far more readily if it were not quite as sour as the original version. After sequestering our friends and plying them with dozens of different recipes, we finally settled on a creamy, mildly tart yogurt cheese with salt, thyme, and minced onions, which we hoped would find favor with the consumers.

"Nobody is going to know what it is, though," I said, feeling more and more worried as our yogurt cheese debut approached. "We should print little cards explaining how it can be used, and maybe add a recipe or two."

I worked feverishly in the next day or two to prepare some material indicating that yogurt cheese could be used as a dip with crackers, potato chips, or raw vegetable sticks, or it could be used as a sandwich spread or a topping for baked potatoes. I gave recipes for cheesecakes, salad dressing, and several different desserts and sauces, emphasizing that yogurt cheese had fewer calories than cream cheese or sour cream and could be used as a substitute for either. I went on to describe the excellent nutritional value of the product, pointing out that unlike other dips it contained no gums, water, fillers, or preservatives. I then presented the copy to Consumer and Corporate Affairs for their approval.

"You'll have to delete the last paragraph," they said firmly. "You're not allowed to talk about what your product *doesn't* contain. That falls under the regulations covering negative claims, so you'll just have to take it out."

I didn't bother to argue with them. I knew only too well that the material would be confiscated if I went ahead and printed it anyway, but I was no closer to understanding what was accomplished by withholding this sort of useful information from the public.

During the Christmas season the yogurt cheese sold surprisingly well. Our large signs and printed cards succeeded admirably in drawing the attention of the customers to our new product, and we cheerfully struggled to keep up with the unexpected demand. But as soon as New Year's Day was over, the bottom suddenly dropped out of the yogurt cheese market. The product had evidently been perceived as something to serve at a party, but nobody, it appeared, was willing to eat it alone.

"Just when we thought we were going to be able to make our drops worthwhile!" I said, feeling thoroughly discouraged. "We're going to have to do something to move that yogurt cheese, now that we've paid for the containers, the plates, the printing, and a mountain of cheesecloth bags."

"Well, if we can't get Nova Scotians to change their eating habits, then let's try to send it somewhere else, where the people are more used to variety and unusual foods."

"Wouldn't it be wonderful if we could sell it in New York," I mused, thinking nostalgically of the many times we had shopped at places like Balducci's when we lived in Greenwich Village. I felt it was time to pay a visit to my new friend Roger Mason at the Nova Scotia Dairy Commission.

"I'm afraid it will be impossible for you to ship your yogurt cheese to New York," he said, smiling sadly. "You see, both Canada and the United States are so over-produced in milk that they've decided to put a full moratorium on it. So that means neither country can import or export milk. I'm sorry."

"Well, I can understand how that would apply to fluid milk or yogurt, but this is *cheese*. How do cheeses like Gouda or provolone or Camembert get into the States, then?"

"The Americans were importing them from Europe before the moratorium agreement, but now the doors are completely closed."

Nevertheless Roger was kind enough to look more carefully into the matter, and sure enough he phoned one morning to say he had found a loophole. The Americans were willing to import up to two and a half million pounds of any cheese they didn't already have or produce themselves, and our yogurt cheese was unique enough to qualify.

The dairy officials in Ottawa were delighted to hear

that the two and a half million–pound quota was going to be filled at last, and they immediately contacted their counterparts in Washington to help smooth the way for us. Everything went beautifully until I received a call from an agent at the Food and Drug Administration who told me I could not call my product *yogurt cheese*. According to U.S. regulations, yogurt must contain a certain percentage of whey, and when it is removed it can no longer be called *yogurt*.

"Of course," I replied, "that's why I'm calling it yogurt *cheese*, because cheese by definition is fermented milk with some of the whey taken out."

The FDA official was sympathetic but unmoved. The regulations stipulated that I *must* call my product *soft unripened skimmed-milk cheese*. The word *soft* had to appear in the title because of the particular moisture content, *unripened* because the product had not been aged, *skimmed-milk* because of the butterfat content, and *cheese* to indicate the removal of whey.

"But that's too much of a mouthful," I objected. "And besides, it's *boring*. Who would ever be tempted to buy a product with such a long, dull name?"

But the official stood firm, bound irrevocably by the regulations he was being paid to enforce. I was told I could not use the word *yogurt* anywhere on the package, lest it mislead the public. The word was banned from the title, the subtitle, the supratitle, and even from the list of ingredients. Just as I was beginning to accept the fact that I had lost the argument, I suddenly remembered I was known affectionately in Nova Scotia as the "Yogurt Queen."

"Would it be all right if I used the words *Yogurt Queen* as my brand name?" I asked breathlessly.

"You may use any brand name you wish, as long as it hasn't already been trademarked."

For two full weeks Gordon and I rejoiced at the thought that our unique yogurt cheese would allow the farmers of Nova Scotia to produce five million extra pounds of milk per year, but Yogurt Queen soft unripened skimmed-milk cheese was not to be. One morning I received a call from an official in another branch of the U.S. government informing me that because my product could not be called yogurt cheese, it no longer qualified for the quota allotted to cheese not made in the United States. There was plenty of soft unripened skimmed-milk cheese everywhere, he told me, and the Americans were not about to import more from abroad. At first I was devastated by this bureaucratic catch-22, but I later found out that even if I *had* finally succeeded in getting my yogurt cheese down to New York, I would have had to pay a 30 percent fee to a licensed importer, another 10 percent to a cheese broker, and still another 30 percent to a distributor. Evidently the export market existed only for companies who could afford to sell their products at a loss.

Taking the funny farm on the road

"I'm afraid we're suffering from poor economies of scale," I said to Gordon one morning, after spending two rather discouraging hours reading our ledger. "Distribution is still our main problem. The trucks are costing a fortune. Our drops are so small they hardly even pay for the diesel fuel, let alone the repair bills and the drivers' salaries and expenses. We're going to have to do something pronto."

As we stood poised on the brink of our eighth year in business, we knew we would have to either cut our costs, raise our prices, or improve our sales. We both agreed that the first two options were out of the question. It would be difficult to cut our costs without affecting the quality of our yogurt, and we viewed a price hike as an unattractive measure to be taken only as a last resort. The only acceptable alternative was to improve our sales.

"It's not going to be easy, though," I pointed out. "We already ship our yogurt to just about every major outlet in the Maritimes now. We can't pick up the corner stores because they're too small to justify a stop. And it's no use putting out another flavor, either. The last one we introduced cannibalized the others."

"It *what?*" said Gordon, looking startled.

"That's a term I picked up the other day," I said, smiling. "It means that people drop one flavor to buy a new one, but the overall sales remain the same."

"Well, our overall sales are *not* going to remain the same," said Gordon with finality. "I found out from the Department of Agriculture that only five percent of the families in Nova Scotia eat yogurt. And that, in my opinion, is downright ridiculous. There's not one single reason why *everybody* shouldn't be eating our yogurt every day of the year. I'm not going to go to the trouble of putting out a product worthy of feeding the gods on Mount Olympus, only to have people pass it by and leave it to expire on the supermarket shelves!"

"What are you going to do about it, then?"

"I'm going to build myself a display case and then I'll put on a white lab coat and a bump cap and I'll go to all the stores and personally see to it that the other ninety-five percent of the people in Nova Scotia try our yogurt."

"But how are you going to get them to try it?" I asked. "You can't exactly force them to put it in their mouths if they don't want it. And you know what people think of yogurt around here."

"We at Peninsula Farm have our ways," said Gordon, rubbing his hands. "I'll set up my booth right in front of the dairy case, and if anyone dares to reach for the wrong brand, I'll catch his wrist in midair. And if they try to sneak past me without stopping, I'll wrestle them to the ground. And if they refuse to open their mouths, I'll pinch their nostrils. I've been patient for eight years now, but when they told me about that ninety-five percent, I decided it was time to go into action. And action is what they're going to get!"

The managers were delighted to hear that Gordon was planning to do what they called "demonstrations" in their stores. They liked to encourage suppliers to give out free samples of their products, for this always pleased the customers at the same time that it spruced up the weekly sales figures. These demonstrations were usually quiet affairs, consisting mainly of an unobtrusive card table on which were placed some product samples guarded by a poorly informed young person in a sitting position. The customers would pick up the sample and pop it casually into their mouths while the demonstrator watched silently in the shadows, too shy or too embarrassed to volunteer any information about the product unless questioned by the consumer.

The managers, however, were not prepared for Gordon. They had never before been approached by the owner of a company to do a demo, and they had never seen anyone address the customers with such spirited enthusiasm. Nobody knew more about yogurt than Gordon, and nobody felt prouder and more committed to his product than he. He would accost the unsuspecting shoppers with such warmth and good-natured determination that very few could resist accepting a sample of Peninsula Farm yogurt, and soon his booth was surrounded by the curious, the dubious, and the bemused.

"We're sampling Peninsula Farm yogurt today, sir," he would call out in his friendly voice. "What flavor would you like this morning? We have strawberry right here. Look at that beautiful color! Here, take a spoon. Just smell that bouquet! Those are real strawberries you have in your mouth. They're fresh, uncooked berries. Do you feel how juicy they are when you bite into them? And they were grown right here in Nova Scotia. We're the only yogurt manufacturers who use local strawberries. Everybody else buys Polish or Mexican berries. But we're trying to create jobs here in Nova Scotia. How did you like that yogurt?"

With very rare exceptions, the customers would express their astonishment at how good the yogurt was, thereby encouraging other shoppers to try some too. They would turn to one another with expressions of wonderment, and soon a small crowd would be exchanging favorable comments about their amazing new discovery. Gordon would work hard to attract new people to the throng at moments like these, for he quickly learned that the customers were far more disposed to try something new if they saw others doing it first, and they were also more likely to believe the opinions of fellow shoppers than those of the demonstrator himself, no matter how well-informed he might be.

After doing a few demonstrations in some of the major stores, Gordon observed that the world could be divided into yogurt lovers, yogurt haters, and the Great Undecided, consisting of those who had never heard of it or were afraid to give it a try. Each group, he claimed, could be identified with reasonable accuracy even from a distance. The yogurt lovers were almost always intelligent-looking women with oversize spectacles and sensible shoes. These, in turn, could be divided into two categories: those who swore by Peninsula Farm yogurt and those who had not yet heard of it. It was difficult for Gordon to believe that there were yogurt lovers running around loose who had still not tried our product, but once they sampled it they were quickly converted and promised to spread the word to their friends. Evidently our usual position on the bottom shelf was indeed preventing potential customers from finding our yogurt, but once they knew where to put their hands on it the problem was solved.

The yogurt haters were usually men, especially those who had tattoos or shopped in their undershirts. The only men who did like yogurt seemed to be students, artists, or intellectuals. The yogurt haters could also be divided into two subcategories: those who had tried yogurt and genuinely hated it, and those who had never tasted it and never intended to as long as they lived.

Gordon was inclined to sympathize with the group that had tried yogurt and hated it.

"I can understand exactly how you feel," he would say. "In an unguarded moment you tasted the wrong brand and you've been put off for life. But I'm here today to offer you the opportunity of trying the *right* brand. What have you got to lose? I'm not going to try to make up your mind for you. It's entirely up to you. But this is the only time I'll be in this particular store doing a demo, so why don't you seize the opportunity to sample our yogurt, free of charge, so you can decide for yourself whether I'm telling you the truth or not. Take my word for it, though. It's the yogurt that yogurt haters *love!*"

Nine times out of ten they would try it, either out of curiosity or to get Gordon off their backs, and of those nine at least seven would be honest and good-natured enough to admit that he was right. They would invariably buy some yogurt and walk away shaking their heads in disbelief, smiling at the thought of what their friends or family would say when they, of all people, produced a container of yogurt and ate it in front of everybody.

Gordon found it predictably difficult to deal with the customers who fell into the category of those who hated yogurt without even trying it. They were, of course, a little on the narrow-minded side, but they expressed their opinions with unwavering self-confidence. Unfortunately they were almost always wrong about everything they said.

"You can't put anything over on *me*," declared one. "That stuff has *bugs* in it."

"That's right," said Gordon proudly. "We never pasteurize our yogurt after we make it, as some manufacturers do to extend the shelf life. All the bacteria in our yogurt are alive, and that's one of the reasons it's good for you."

"Don't give me that. You can't tell me germs are good for me. Why have they developed antibiotics, then? All the doctors are trying to kill germs, and you're standing there telling me they're *good* for me."

"But there's such a thing as beneficial bacteria. They're not all pathogenic . . ."

"Look, you told me yourself you don't pasteurize your yogurt. That's got to be illegal. You'll give us all tuberculosis."

"We pasteurize the *milk* before we make the yogurt. Once the milk is pasteurized, there's no danger at all."

"Oh yeah? You say you pasteurize the milk, but what's that yellow stuff on top of the yogurt? It's cream, right? I know cream when I see it. I wasn't born yesterday. So that *proves* you don't pasteurize your milk. I ought to report you to the authorities!"

"We *do* pasteurize our milk, but we don't *homogenize* it," Gordon explained, as patiently as he could.

"Same thing," said the customer, stalking off in the direction of the manager's office while the other shoppers murmured suspiciously among themselves.

Such frustrating conversations happened infrequently, however. Most of the customers were undecided about whether or not they wanted to try yogurt, but Gordon was usually able to convince them to take the plunge. He was well prepared for them as they approached his booth, for he could categorize them in advance as Undecided by peering into their shopping carts, where there would be generous quantities of pop, chips, candy, and assorted junk food. The idea of eating something as exotic as yogurt had simply never occurred to them.

"I don't even like the sound of the *name*," they would say.

"Which syllable do you dislike the most, 'yo' or 'gurt'?" Gordon would counter.

"Both."

"Would you give up eating steak if it were called 'blech' or 'yuck'?"

"Probably."

"Then just pretend this product is called 'creamy shortcake filling,' and give it a try."

The excuses for not trying our yogurt were legion, but Gordon persevered. By dint of sheer tenacity, humor, and irrepressible energy, he managed to conquer, over the course of the year, most of the doubt, fear, and misinformation that lurked in the minds and hearts of the province's Reticent Majority. When our accountants handed us our year-end financial statement, we were delighted to find that Gordon had single-handedly doubled our annual sales.

Farmer Jones hits the talk shows

Very few inhabitants of Nova Scotia escaped an encounter with Gordon at some time during his weekly performances in the stores. Among those who met him were various writers, journalists, and radio or TV producers who recognized immediately that the dynamic demonstrator in the white lab coat was no ordinary salesman. This was a crusader, and people were beginning to join the "cause" with all the dedication and zeal of true believers.

"Mr. Jones," said one talk-show host, "obviously you're doing well because your yogurt is the best. We might as well call a spade a spade. But what I'd like to know is why can't other dairies do the same thing? Why are really delicious, world-class products getting harder and harder to find on the market today? Whatever happened to excellence?"

"Excellence, I think, is being killed by the bottom line. That's the way I see it, anyway. For one thing it's getting just about impossible to do business anywhere in the world today without paying somebody off just to be allowed to exist in the marketplace. The buyer holds the purse strings, so of course he gets to call the shots. He wants cheap goods that can be sold at a high profit margin, and on top of that he wants the best deal he can get for doing business. This puts a tremendous burden on the supplier, who generally ends up cutting costs in order to be competitive. But being competitive doesn't mean being excellent anymore. It means having the funds to offer the best deal."

"So how does this affect you? Have you ever been turned down for not being competitive?"

"I certainly have. I can't begin to tell you how many bids I've lost with institutions who put out tenders for yogurt. It's their policy to buy the cheapest product, but quality seems to be of no concern to them at all."

"That sort of thing must happen across the board. Stadiums cave in, bridges collapse, high-rises burn down. I'll bet your bottom line is the culprit somewhere along the way. So do you think excellence is doomed?"

"I hope not. I think a lot will depend on the consumer. That's why I've been out there talking to people in the stores. I wanted them to know that our company is committed to excellence, and I wanted them to taste our yogurt so they could see for themselves what kind of performance they can expect from a food manufacturer who has made the decision to put out the best product he possibly can."

"And how did they react?"

"They were excited, most of them. They never knew yogurt could be so good, and then they began asking how they could put the pressure on other manufacturers to do the same thing with *their* products."

"And what did you advise them?"

"I told them to exercise their purchasing power. If they buy only the best and refuse to buy schlock, after a while the substandard items will have to disappear."

"Did they tell you they couldn't afford to buy the best?"

"Yes, but those arguments don't hold water, of course. When you sit down and analyze the nutritional value of junk food, you soon find out it's not worth the package it's wrapped in. It costs more by far to try to feed a growing family on junk food than it does to go out and get something decent to eat."

"So you're depending on the average consumer to have a pretty sophisticated palate."

"Well, that might be asking a lot. Don't think I won over *all* the customers in my store demos. There's a yogurt on the market that's just loaded with artificial color and flavor. It tastes like lollipops, but some of the shoppers insisted they liked that brand better than ours. So there are actually people around who prefer artificial flavor to the taste of real fruit, for instance. What can you do? You can't win 'em all, but there are enough customers with discerning taste out there to really make a difference."

"I hope you're right," said the smiling host as he wrapped up the interview. "And I know what I'm going to do first thing tomorrow morning. I'm going right out and buy some of that delicious yogurt of yours. Excellence is beautiful, and we all ought to support it."

14

The Battle Continues

Goliath brings out the guns

Gordon's demonstrations and his subsequent radio and television interviews had such a beneficial effect on the business that we decided to award ourselves a salary for the first time in our long struggle. It was a pleasure for us to know that our work had not been in vain. We attacked the grueling warm-weather chores with renewed spirit and energy, and even Daisy the Cow, who by this time was getting on in years, was as frisky as a heifer in the newly greened meadows.

Ed Shoemaker, on the other hand, was apparently unaffected by the arrival of spring, for he chose the bluest, sunniest day of the season to summon me to his office to discuss the development of a "plan" between Peninsula Farm and Goliath Food. Once again I was left for about an hour to cool my heels in the outer office, while the receptionist, who had as yet found nothing to relieve the painful tedium of her existence, busied herself with the task of doing her fellow mortals the favor of answering their telephone calls.

"Everybody has a plan," said Mr. Shoemaker, when at last I had settled myself in the chair next to his desk.

"I'm sure I don't need to tell you that we work on volume discounts here. Your volume has been going up steadily for the last couple of years, so it's time now for us to talk about what kind of a discount plan you can offer us."

"I'm sorry," I began, feeling puzzled. "I don't know where I've been lately, but somehow I never realized you had central warehousing for refrigerated products."

"We don't," said Mr. Shoemaker tersely.

"You'll have to excuse me, then. I may not know that much about business, but to me a volume discount is a deduction made by a supplier when he can cut costs by selling goods in large quantities. But I don't quite see how this applies to our situation. You have stores all over the province, and I have to send my trucks to every single one of them. It doesn't make any difference to me whether they're all owned by Goliath or they operate independently; it still costs me the same to deliver the yogurt. If I were selling you a carload of canned goods and dropping them off at a warehouse, I'd be happy to pass the savings on to you. But in my case I'm not saving anything at all."

"I know what you're saying, but it just doesn't work that way. Everyone else gives us a volume discount, so you're going to have to kick in something too. It's as simple as that. When you were small and just starting out, I could look the other way, but now you're getting to be a real factor in the market. You're going to have to be more competitive, or why should I have you in our stores?"

"Why? So you can give your customers what they want! What's wrong with that?"

"Plenty. The more yogurt you move, the less the other dairies can sell. And when their sales go down, so do my rebates from them. It doesn't make much sense for me to keep you on the shelf unless you can match their deals."

"But what about the *customers?* Don't they have any say? Aren't you afraid they'll be disappointed if my product disappears from your stores?"

"Yogurt is yogurt, Mrs. Jones. Your brand is not the only one people buy. And your product is certainly not the only item we stock. I'm sure Goliath would have no trouble at all surviving without your yogurt. But can you say the same thing yourself?" he added, smiling significantly.

I stared at him indignantly. "Your customers will go to the other chains to buy our yogurt. And once they're inside the stores, they'll get the rest of their groceries, too."

"You win some, you lose some," he shrugged. "But Goliath is known for its low prices, so they'll be back. People do a lot of comparison shopping, you know, and they go for the bargains. And we like to see to it that they get them. Do you follow me? So think it over, and when you've worked out a plan, come back and see me."

"You can't get blood out of a stone," I said, trying to hide my outrage with an unnaturally pleasant smile. "The dairy business is unbelievably capital-intensive. We've had to plow all the profits into tractor-trailers and stainless steel equipment. Everything we earn goes straight into growth and expansion, so by the end of the year there's practically nothing left."

"We all have our problems," said Mr. Shoemaker, leading me to the door. "But I have to be fair to the other dairies too. It wouldn't be right for me to take more from them than I get from you, now would it?"

By the time I got home that evening, Gordon was already in the barn milking the cows. The air was warm and smelled of short-feed and manure. The sounds of rattling stanchions and the rhythmic pulsations of the milking machine competed innocently with the six o'clock news. Vicki was giving the calves their bottles; Valerie was dumping pails of expired yogurt over the heads and backs of the pigs who were standing expectantly in their feed troughs; and Billy, our newest cowhand, was loading the wheelbarrow and pushing it along the plank to the manure pile. Gone were the days when Gordon would do the barnwork alone while I made yogurt and wrote books on the kitchen stove. We had help now, and the nightmare of endless hours of slave labor was promising to become just a distant memory. We had succeeded, almost against the odds, in dragging ourselves through the first part of the tunnel, but the light appeared to be blocked out entirely by the threatening figure of Ed Shoemaker and company.

"It's like the protection racket," said Gordon later that evening. "What a laugh. He claims he's being fair to everyone by demanding the same rake-off, but all we're really doing is buying permission to be in the stores."

"And just when we were about to take a salary," I added bitterly. "You know where *that's* going. Right into Ed Shoemaker's pocket. Why should he be getting the profit from our labor, anyway? He hasn't done a thing to deserve it."

"What gets *me* is the image Goliath has made for itself as a low-cost chain that keeps its prices down by being efficient and taking only a small profit margin. But nobody realizes they're getting most of their profit from the *suppliers*. And what's worse, they stock all their stores with the products that give them the best deal rather than the ones with the highest quality. I thought people could exercise their purchasing power by refusing to buy schlock, but they can't very well buy the good stuff if it isn't even on the shelves. You may as well say good-bye to quality and freedom of choice, too!"

"So what shall we do? Should I try to call Stuart Ritchie again and cry on his shoulder?"

"I'm afraid he's not going to have time for a little company like ours. We'll have to figure out how to survive on our own. We can't depend on other people to bail us out, anyway. Those stopgap measures aren't really sound in the long run."

But before we had a chance to plan a strategy, we were suddenly deluged with the problems and consequences of a new crisis, provoked by none other than Ed Shoemaker himself. In his wisdom he had seen fit to raise our retail prices from $1.59 to $2.09 for a 500-gram container. The consumers found the 32 percent increase distinctly unappealing, and our sales tumbled to roughly half our normal volume. Even during the worst yogurt wars there had never been such a dramatic spread between the price of our yogurt and that of our competitors. What was worse, our customers assumed that *we* were responsible for the price hike, so Gordon was kept busy throughout the following days answering phone calls and sending out letters of explanation to the many irate shoppers who contacted us.

Meanwhile I jumped into the car and drove straight to the head office of Goliath Food. As I had no appointment I was kept waiting for the better part of the morning, but finally Mr. Shoemaker agreed to grant me a moment of his time.

"It's amazing how quickly the suppliers show up around here to talk deals when there's a change of price in the stores," he smiled, complacently lighting a fat, mud-colored cigar.

I felt demeaned and furious. "Is that why you raised my prices? Just to get me in here to talk deals?" I asked sharply.

"Now there's no need to get all excited," he said serenely, as he puffed at his cigar to get it going.

"We've lost several thousand dollars in sales already

this week," I said, making an effort to be agreeable. "I can't help feeling upset."

"I can understand how you feel, Mrs. Jones. But try to put yourself in my position. If you don't send anything here to head office, then I have to make up the difference by getting it from the stores. I don't like to do that, because I want to maintain our low-cost image, but you leave me no choice. I tell you what I'll do, though. You work out a plan for me, and I'll put your prices back down where they were. That way we'll both benefit. Your sales will go up and I'll have something to show for carrying your product."

"Isn't there some sort of a law about that? This feels like a shakedown to me."

Suddenly Mr. Shoemaker lost his expression of amicable geniality. He sat straight up in his chair and glared at me with narrowed eyes.

"Let's not get funny, Mrs. Jones. Everything we do at Goliath is well within the law. Just what are you trying to imply, anyway? I think it's about time we got something straight. We're not asking for *kickbacks*, you know. We don't use that term around here. What I'm talking about is an earned cost reduction. We call it an ECR. You'll by paying us for services rendered."

"This is the first time you ever mentioned any services," I said, shifting my position on the chair. "What will you be doing for me, exactly?"

"Cooperative advertising, sales, that sort of thing," he said vaguely, glancing nervously at his watch. "Look, I have a meeting to go to now, so we'll have to wrap this up. Your prices go down when we have a deal. It's up to you. Just let me know."

Ed Shoemaker had me over a barrel, and I hated it. I was well aware that my customers, even the most loyal ones, could not afford to pay fifty cents extra for a tub of yogurt. I also knew Peninsula Farm could never survive without the Goliath outlets, for they represented the largest portion of our market. I realized that if I didn't move quickly to get Goliath to lower the price of our product, the other stores would soon raise their tags to narrow the gap and thereby increase their own margins. It would most certainly be the beginning of the end, I thought unhappily, if our sales were to suffer as drastically in all the stores as they had at Goliath Food. But pay a kickback to that cigar-chomping stereotype of a hard-nosed businessman? Never.

I decided it was high time I paid a visit to David Sobey for a heart-to-heart talk about the dynamics of the buyer-seller relationship in the world of commerce. I had spoken to Gordon on a number of occasions about going to Sobey for advice, but he had never felt very enthusiastic about turning to others for help or counsel. Maybe this was the New Yorker in him (he would never even *consider* asking directions from passersby lest they suspect him of being a potential pickpocket), or perhaps as a man he felt he should fight his battles in grim solitude; but as a woman I was comfortable with the idea of going to an experienced and well-informed professional and asking for his help in understanding the nuances of doing business in the Maritimes. He was, after all, like a godfather to Peninsula Farm, for if he hadn't tracked us down on that fateful day so many years ago and asked us to supply his supermarket chain with our homemade yogurt, we might never have become an entity in the Maritime economy. I was sure he had never intended to encourage our growth and development for the purpose of having us end up lining the pockets of the Ed Shoemakers of the world.

"You accused Eddie Shoemaker of shaking you down?" said David Sobey, leaning back in his chair and contemplating me with a barely suppressed smile. "I'd give anything to have seen his face," he went on, allowing his smile to broaden. "I don't think he's used to having suppliers talk to him that way."

"But I was right, wasn't I?" I said, hoping to be vindicated.

Sobey looked at me thoughtfully for a while before shaking his head. "No, I can't say you were right, I'm afraid. But I can certainly understand how you felt and why you believed you were being muscled by this man. Eddie Shoemaker doesn't have a reputation for subtlety. He fought his way up from the bottom, and he's still a little rough around the edges."

"But it still amounts to the same thing, doesn't it? He wanted me to make him a payment so I could keep his account. He didn't exactly say he'd throw me out if I didn't pay him, but he raised my prices so high that nobody's going to buy my yogurt if I don't knuckle under to him. He claims he's going to do something to earn the kickback, like advertise my yogurt maybe, but I think it's just an excuse. It's a scam, if you ask me."

"I know how you feel," said Sobey soothingly. "I can understand your indignation, but there's another side to it that I don't think you've had an opportunity to examine. It was actually the dairies themselves who introduced the idea of making payments to head office. You see, in their effort to compete successfully on the market they were anxious to lower the retail prices of their by-products in the stores, and we agreed to do this as long as they made up the difference in the rebates. So it's really the *consumer* who benefits from this arrangement, and I don't think any change in this long-standing plan would be very popular with the customers, because what's happening is that the suppliers are contributing part of what the shoppers would have to pay for any given item. But don't think of it as a kickback. There's really nothing nefarious about it at all. The spread between the wholesale and the retail price remains the same, but by paying some of it directly to head office you're offering your customers an excellent product at an attractive price. I'm sure you're already well aware of the need to keep your volume high in order to make your business profitable."

"All I know," I said cautiously, "is that if I start paying rebates now, I don't see how I'll end up with any profit at all."

"I understand. You're a young company, and you're still struggling to get on your feet. As far as I'm concerned, I'm willing to wait as long as it takes to let a new local industry get off the ground. But you should realize that you're being given special treatment because we all want to encourage local business. It benefits everybody. Eddie Shoemaker has jumped the gun a little bit, that's all. But he's not going to throw you out of Goliath, and he's not going to put you out of business, either. He may be crude, but he's not stupid. You'll probably find he'll be quite lenient with you until you catch up with the other suppliers. Just make sure that when you work out a plan with him you offer the same rebate to all the chains. That's an unwritten law in the food business, and if you break it by giving one chain a better deal than the others, you'll find yourself delisted and you won't get much sympathy from anyone. Otherwise, you're going to be just fine. You make the best yogurt I've ever tasted, and I'm proud it's produced here in Nova Scotia."

By the end of the day I had spoken with the managers of several other dairies to get their side of the story, and I found that Sobey's assessment of the situation was quite accurate. The dairies had indeed offered to give rebates on by-products in order to be more competitive, but as time went by they had outcompeted one another to the point of losing more profit than they had anticipated. They tended to blame the chains for pressuring them for higher rebates, but they also grumbled at one another for trying too hard to grab extra market share. My friends at the dairy commission corroborated their views, but they also emphasized that they had only themselves to blame for their low profitability. The commission had originally passed a regulation stating that fluid milk was exempt from rebates, but one by one the dairies had ignored this protective measure and started rebating milk as well. When the red ink started to flow, they begged the commission to put some teeth into the regulation and monitor it more strictly, but by that time it was too late. The commission had already decided to dismantle the regulation so as not to be forced to prosecute the dairies.

I finally came away with the impression that the business world was really just a complex and highly evolved Turkish bazaar, with everybody arguing and bargaining to find a middle ground where all parties would end up feeling reasonably satisfied with their deal. The relationship between consumer, buyer, and producer reminded me of a game we used to play as children in which a rock could crush scissors, but scissors could cut paper, and paper in turn could wrap rocks. Usually each child had a similar amount of power and vulnerability, but if things got out of control it led to tears and fistfights. My day with Sobey, the dairies, and the dairy commission convinced me that nothing had changed in the adult world. It was still a game of scissors, paper, and rocks, but it seemed to be self-monitored with reasonable equitability. Nobody, as far as I could tell, was walking off with excessive profits, and the consumer was paying a fair price for his food.

The most important lesson of the day for me was the realization that Peninsula Farm's success was partly due to our having been treated with patience and leniency by the chains. I had always taken it for granted that the credit belonged entirely to us because we had worked extra hard and had put out a world-class product. But if we had been required to pay rebates from day one, the financial strain would have been an impossible burden. I was equally grateful to the managers of Twin Cities Dairy for having sold us milk when our small herd of cows proved unequal to the task of keeping up with the demand for yogurt. Although they made yogurt themselves and were directly competing with us for shelf space in the stores, they generously sold us all the extra milk we needed because it benefited the farmers who belonged to their cooperative. It finally dawned on me that I was living in a business society where the idea of "all for one and one for all" seemed to work. There were arguments and recriminations, of course, and the usual amount of pushing and shoving, but the overall perception seemed to be that if local businesses could not directly help one another they should at least refrain from grinding the newer ones into the dust, for a strong local economy provided much-needed employment and opportunity for growth. The government also reflected this attitude, for it had given us various grants and loans when we most needed assistance. Federal and provincial representatives on all levels were well aware that successful business start-ups would ultimately serve to strengthen the local economy as a whole.

If my sessions with my counterparts in the business world taught me a lesson in moderation, they also helped me unravel the meaning of those two most troublesome words, *kickback* and *rebate*. Webster's definitions helped me not at all, but it became clear to me that day that a *kickback* refers to a payment made against one's will to an unappealing individual (probably a yogurt hater) who appears to have one up against the wall. A *rebate*, on the other hand, is a portion of the purchase price directed back to the buyer in consideration of the well-being of the consumers.

With our new understanding of the rebate arrangement firmly in mind, Gordon and I contacted the chains with a plan we felt we could handle fairly well at that stage in our development. Although it dropped our potential profit right back down to break-even again, we were confident that if our growth continued as it had in the past we would eventually be able to enjoy a decent return on our labor and investment. It would simply have to be put off once again until a later date.

More new products

"Our profit margin is so low now that we'll have to go all out to improve our sales just to stay afloat in this market," said Gordon, looking up from the ledger. "It's getting serious now."

"So what shall we do for our next trick?" I asked.

"Invent something new," said Gordon brightly. "I'm the market-research department, and my statistics tell me that the per capita consumption of yogurt in Nova Scotia is still well below that of many other areas of the world. We don't have time to wait for everybody to discover yogurt here, not when we have to pay rebates every three months. So let's give them what they want."

"Well, since I'm the product-development department, I'll need to know what your data tells us about consumer preferences. What sort of a product do you think would have universal appeal?"

"It has to be cheap, sweet, and low in calories. Preferably chocolate-flavored."

I spent the following weekend laboring away at the kitchen stove trying to produce a pudding made from skimmed milk and aspartame (an artificial sweetener), but nobody was in the least bit impressed but the pigs. The mixtures were so watery and unappealing that I finally gave up my calorie-consciousness and started experimenting with creamy, rich concoctions that were quickly devoured by Gordon and the girls before they even had a chance to cool. The final recipe involved milk, cream, butter, pure cocoa, and our favorite chocolate bars. Although most puddings call for eggs, I decided these would have to be eliminated altogether because of the possibility of their causing food poisoning if the product were to be exposed to lengthy bouts of unnecessary warmth on loading docks, in car trunks, or in poorly functioning coolers.

"Gordon, I've done everything wrong," I admitted, as we gorged ourselves on the results of experiment number sixteen. "This pudding isn't cheap, and it's not low in calories, either."

"Yes, but it's *good!*" exclaimed Vicki, making no attempt to wipe off her chocolate moustache.

"Well, there you are," laughed Gordon. "The ayes have it. We'll just have to come to terms with the fact that a really first-class product can't be made with chemicals and watered-down milk. We can't be all things to all people."

"We're going to have the same problems, though. The other puddings will be less expensive, so of course they'll sell like hotcakes. And who knows what the rebate on pudding is. It'll probably be sky-high, and we'll never be able to handle it."

"Don't start getting pessimistic now," chided Gordon. "Quality will get us through, somehow. There are people out there who are just as committed to Peninsula Farm as we are to excellence. They buy our yogurt because they know they can depend on it to satisfy them, but they have to be given a little variety once in a while. We should devote more time to developing new products.

Nobody can eat the same old thing all the time, no matter how good it is."

"That's true, Mum," said Valerie, scraping the last of the chocolate pudding out of the double boiler. "I'm tired of our yogurt flavors. Why don't you make blueberry for a change?"

"You should, you know," Gordon agreed. "Just the other day I was contacted by someone from the Department of Agriculture who wanted us to put out blueberry yogurt. There's a glut of wild blueberries on the market right now, and the Blueberry Association of Canada is going crazy trying to get them sold. They're shipping them all the way to Japan, did you know that? It seems a shame for them to have to send them that far when they have a buyer like us right here in their backyard!"

The development of our wild blueberry recipe, like that of all the other flavors we had on the market, took approximately half an hour. I simply dumped some freshly thawed frozen berries into a bowl of plain yogurt and added sugar to taste. The results were superb. The blueberries burst in our mouths as we devoured the contents of the mixing bowl, and within minutes Gordon was on the phone with our graphic designer, ordering new containers.

"One thing that's good about being small," he remarked, rubbing his hands with satisfaction, "is that you can get things done *quickly*. In a big company it would have taken months, or maybe even years, to come out with a new product. But what I like to see is *action*."

It only took the graphic designer one week to do the artwork, and the printers had the plates ready in record time. Within a month a tractor-trailer was maneuvering to back up to our loading dock with the first shipment of new containers, and five days later our own trucks were pulling out of the farm, laden with pallets piled high with

Peninsula Farm wild blueberry yogurt. Gordon and I watched happily as the drivers made their way down First Peninsula.

"We should have done this long ago," said Gordon, giving a thumbs-up sign to the last truck. "People around here *love* blueberries. It'll probably be our most successful flavor."

Our wild blueberry yogurt launch turned out to be an unmitigated disaster. Three days later the phone was ringing off the hook with complaints from customers and store managers from every corner of the Atlantic provinces.

"What kind of yogurt are you shipping us, anyway?" demanded an angry manager from a store in Newfoundland. "The stuff's all runny, just like milk. I t'ought you fellows put out a good product, but I'm not paying for *this* shipment, you can be sure o' that!"

I dashed to the nearest store in Lunenburg and bought a container of our wild blueberry yogurt. When I lifted the lid, I discovered to my horror that the creamy yogurt had indeed turned into purple milk, with tough little dried-out skins floating on the top. The yogurt had apparently leached all the juice out of the berries, and the resulting mess was enough to dismay even the most sympathetic customers.

"It's my fault, I'm afraid," I groaned. "I ate all the batches as fast as I mixed them, and the rest of you gobbled up whatever was left over. It never occurred to me to leave them in the fridge for a couple of weeks to see how they'd behave. The fresh frozen strawberries and raspberries never pulled a stunt like that, so I took it for granted the blueberries would be the same."

"Well, I'm just as much to blame," said Gordon. "Your recipe was so good I couldn't wait to get it out on the market. I was sure the customers would go wild for our blueberries. But not *this* kind of wild!"

Gordon spent the next two days making arrangements to get all the unwanted milk shakes off the shelves as quickly as possible, while I sought the advice of food scientists at the Nova Scotia Research Foundation.

"It's a question of reverse osmosis," I was told. "The juice gets out through the membranes of the raw berries, but the yogurt is too thick to pass through the other way. You'll have to cook the berries to arrest the process."

I carefully made up a chart of times and temperatures and spent the next month or so doggedly trying to pinpoint the perfect combination for blueberries that would be not only tender and flavorful but also endure three weeks without letting out their juice. The results were not altogether satisfactory. The blueberries were not as tender as they were when they were freshly thawed, and the flavor seemed to be somewhat altered in the cooking.

My next experiment involved slitting the raw frozen berries with a razor blade to see if by making a sizable opening in the skins I could equalize the entry and exit of juice and yogurt, but all to no avail. I only succeeded in cutting my finger and dyeing myself purple, but I was unable to stop the berries from once again liquefying the yogurt around them.

It finally dawned on me that if fresh frozen blueberries let their juice *out*, then surely freeze-dried berries would have to suck the moisture from the yogurt *in*. Gordon located a source of freeze-dried berries in Oregon, and to our great delight we ended up with an absolutely delicious batch of blueberry yogurt. The only problem was that it took the berries about eighteen hours to rehydrate in the yogurt, and the mixture had to be stirred afterward to eliminate an otherwise splotchy appearance that might have been considered objectionable. This meant that the blueberry yogurt would have to remain for a day and a half in a mixing vat before it could be processed, thus causing endless problems related to production scheduling and contamination by heat exposure.

"I think you're on the right track, though," said Gordon enthusiastically. "What we need are some freeze-dried berries that will rehydrate *quickly*, and I think I know just where to find them. I was talking to a chap the other day who was telling me that the Irish are on the cutting edge of the freeze-dry industry, and they've made some tremendous strides with puff-explosion. You know puffed wheat, and puffed rice? Well, the Irish are puff-exploding everything now. Peas, beans, berries, anything that's more or less round. They take it and pow! You've got a blueberry with a hollow center, just waiting to be filled with yogurt."

We decided to phone the owner of the puff-explosion factory in Ireland. He was a hearty, outgoing man with a booming brogue that sent sonorous vibrations along the transatlantic cable. He was excited, he told us, by our ideas about a new application for puffed fruit, but unfortunately he couldn't get his hands on any blueberries because they didn't grow in Ireland. Would we like him to send us some puffed black corns instead? These sounded less than appetizing to us until we realized he was referring to black currants, but we decided it was best to stick with the blueberries. There was nothing for it but to fly to Ireland with a thirty-pound case of frozen blueberries so that our new friend could explode them for us.

"There's a filling machine in Denmark I wanted to see anyway," said Gordon. "We can't go on dipping the small sizes by hand. It's insane. If we're ever going to grow enough to pay those rebates without hurting too much, we'll have to start by getting a filling machine that can handle the small containers for us."

"Well, if we're going to Ireland and Denmark," I said, "we may as well take a couple of side trips to Hol-

land and Switzerland and track down the perfect chocolate for the perfect pudding. I still haven't given up on
them, you know. Maybe if I can make them irresistible
enough they'll help to keep us in business."

We were still convinced that excellence would save
us from the ever-present danger of the bottom-line philosophy that seemed so prevalent in the business world,
yet there were times when I wondered at what point we
would outgrow the small but distinguished group of customers who were willing to pay for quality and had the
sense to recognize it when they saw it. I was often fright-
ened by the thought that this group was greatly out-
numbered by those who could easily be persuaded to
succumb to the schlock-mongers who had the where-
withal not only to offer the best deals but also to buy the
slickest ad campaigns and the flashiest propaganda. As
we winged our way over the Atlantic toward the Emerald
Isle, I knew that if Peninsula Farm was to prosper we
would have to strengthen our dedication to uncompro-
mised quality with renewed faith in the ultimate edu-
cability of even the least discerning palates of the
population at large.

A view from Europe

We found our frozen blueberries waiting for us in the
baggage-claim area at the Dublin airport, spread out in
every direction on an otherwise immaculate floor. The
package had been crushed beyond recognition by an unknown object of the approximate weight of a grand piano
or a smallish elephant, but when we arrived on the scene
the offending item was nowhere in sight. In perfect evi-
dence, however, was a group of neatly uniformed airline
employees contemplating with obvious distaste the pro-
fusion of slightly thawed blueberries that lay in reckless
abandon at the feet of the incoming passengers. But the

proportions of the tragedy were not so great as we had originally assumed, for we were able to rescue several pounds of product from the crushed box while a team of maintenance men speedily mopped up the more wayward berries before the warm day could turn them into purple puddles.

After being lost for about half an hour in the parking lot reserved for rented cars, we finally sallied forth in the direction of County Cork. Gordon drove cautiously along the left-hand side of the road while Vicki and Valerie drank in the new scenery from the back seat of our Austin Minor. We found ourselves in a country of green, rolling hills, covered with the most amazingly straight cement fences any of us had ever seen. The Irish were obviously unacquainted with frost heaves or the need to wield fence mallets every spring.

We spent four days in Ireland, during which time Gordon and I donned earmuffs and watched blueberries explode while the girls roamed the turrets and parapets of our Best Western castle hotel. Although the puffed berries rehydrated considerably faster than their freeze-dried counterparts, we found to our dismay that the explosions created a sort of shrapnel that was unpleasantly gritty once the berries were immersed in yogurt. It was hard for us to abandon what had seemed to be our last hope of producing a perfect blueberry yogurt, but the food technologist at the plant assured us that cooking the berries with sugar would yield a preparation tender and durable enough to meet our standards. We toyed briefly with the idea of marketing puffed fruit at popcorn stands, but we came back down to earth when our patient hosts described the costs involved in such a scheme.

Our next stop was Copenhagen, where we checked into a small hotel overlooking a canal crowded with a mixture of modern yachts and heavy, old-fashioned wooden boats. We all felt immediately at home in this setting, where land and sea merged in a city of colorful buildings, gourmet restaurants, and bustling street life. We were delighted to find that dairy products abounded. Here was a nation where yogurt was appreciated and savored at every opportunity by a population whose per capita consumption approached thirty-five pounds a year. We soon had the refrigerator in our kitchenette packed with every available brand on the market, but we were sorry to find that artificial colors and flavorings were rife even in a country as discriminating as Denmark. The yogurt was not unlike the brands to be found anywhere in North America, where milk solids and "natural" flavor enhancers were commonplace.

Our disappointment in the quality of Danish yogurt was soon erased by a demonstration of the filling machine we had wanted to see. The manufacturer was given permission to let us observe it operating in what was described as one of Denmark's smallest dairies. The shortest vats in the establishment were three stories high. Gordon and I had to climb a thirty-foot ladder just to take a peek at the vast quantity of yogurt incubating inside, but the filling machine was nevertheless easily able to process the contents during the course of the day. We watched in awe as the machine quietly filled and lidded thousands of containers with no spills, no smashed lids, no breakdowns, and no fits of temperament. A filling machine of the size we needed would have to be specially made for us, but before we left Denmark we had given the manufacturer a down payment on a unit that turned out to be twenty-five thousand dollars less costly than any we had seen before. Our visit to Copenhagen had more than paid for our trip to Europe.

Next on the agenda was the search for the perfect chocolate. We rented a car in Amsterdam and set off for

Switzerland, leaving a trail of chocolate wrappers and yogurt containers in the garbage cans of every town and village on our route. By the time we reached Zurich we were ready for a steady diet of salad and mineral water. Our odyssey had reconfirmed our preconceived notions that the Swiss make the most delicious chocolate in the world and that too much of it makes you feel sick. But it wasn't until we arrived back in Holland that we discovered a brand of chocolate that left us feeling amazingly well, even after we had devoured considerably more than what we had come to acknowledge as a safe level of consumption. A careful scrutiny of the list of ingredients revealed that *vanillin* (artificial vanilla flavor) was missing. It seemed plausible to us that the presence of this chemical in chocolate might partially account for the queasiness experienced by those who overindulge. A taste test of other products containing the ingredient later appeared to substantiate our hypothesis, and further research also indicated that vanillin is a known carcinogen. Since Droste, to our knowledge, was the only chocolate manufacturer who did not always use this additive, and because their product was superlative in both consistency and flavor, we all agreed to add their vanillin-free milk chocolate to the pure cocoa base in our forthcoming puddings.

Holland also won our vote for the best yogurt in Europe. We did find an excellent hazelnut blend manufactured by a Swiss company, and we ran across some superior brands put out by a few small dairies in France, but the Dutch used a different process that we found appealing. Judging from the mild flavor and gluey consistency, the producers were encouraging the proliferation of *Lactobacillus bulgaricus* by incubating the product for long periods of time at low temperatures. This approach yielded a yogurt with a "slimy mouthfeel" (as it

was described by a Dutch acquaintance) that we all enjoyed, but we seriously doubted that it would gain ac-
ceptance in the North American market. We gave the
Dutch full credit, however, for using fresh milk and for
balancing the bacteria to achieve an unusual effect.

We returned home with renewed confidence in the
quality of our yogurt, which we now knew could compete
favorably with the best of the European brands. We also
had proof positive that there was not another yogurt
anywhere that could begin to compare with our fruit
blends, for even in Holland the dairies used preparations
made from jam or compote. Nobody, apparently, was
willing to face the difficulties involved in handling fruit
that was as close as possible to its natural state, for it
was well known in the industry that the costs would be
high and the risk of spoilage even higher.

These factors were certainly of concern to us too.
We had not forgotten the calamity in Mirage when our
yogurt had fermented and burst in all the stores; nor
could we ignore the constant frustration of dealing with
buyers and competitors whose only concern was the bottom line. Maybe the other dairies knew more than we
did about succeeding in business. There was no doubt at
all that they had the means to sweeten a deal and to
launch a million-dollar ad campaign. But if success can
be measured by quality rather than by power, size, and
profitability, our trip to Europe proved to us that we
were at the front of the line. It was comforting to know
that the best yogurt we could use as a standard for constant improvement was Peninsula Farm.

15

Are We Winning or Losing?

Looking at the downside

As we coasted down our driveway, happy to be home again after our two-week tour of Europe, we were greeted by a disturbing sight. The lights were still blazing in the yogurt plant at ten-thirty at night, and white-clad figures could be seen through the windows attending to the vats, the pumps, and the filling machine. We could see by the cars parked outside the factory that we had a full complement of employees working inside. Batch failure! My mind raced ahead, calculating how long it would take to dispose of two thousand liters of spoiled yogurt. Some could be absorbed by the septic system, some would be donated to the neighboring pig farm, and the rest would have to be taken to the dump. It would all have to be containerized in three-gallon plastic buckets with snap-on lids before it could be moved. We would have to send everyone home to get some rest before starting work the following day. Gordon and I would be up all night.

"Don't worry, the yogurt is fine," said our production supervisor. "But the lidder is missing from the filling machine, and we've had to lid the yogurt by hand for five days now. It takes us till about midnight every night, but we're getting the job done! I don't know how long our hands will hold out, though. We're all getting calluses!"

The employees were amazingly cheerful considering the late hour and the frustrating task of lidding close to six thousand containers of yogurt by hand. But the people in the plant that night were the tested and true, the ones who had been loyal when the going was tough, and they were more like family than staff. I was proud of all of them. Most had been with us for years, and it was a joy to think I wasn't going to have to do any more culling as far as this group was concerned. The last time we had had to fire an employee was when Mitch Carson had been caught cheating on his expense account.

It hadn't been easy to fire Mitch. He had come to work for us when he was just a teenager, and over the years Gordon had treated him almost as a son. We had high hopes for him. He was a steady worker, eager to improve, concerned about detail, interested in learning all aspects of the business, and enthusiastic about contributing his best talents to helping the company prosper. After three years we finally decided we were ready to give him an important promotion. He had passed the test of time, and our recognition of his efforts was long overdue.

We were not surprised when his co-workers seemed less than enthusiastic about our plan to train him for a supervisory position. Mitch was the youngest member of the team, and we expected some resentment from the others. But when we questioned them about their objections, our queries were met with uncharacteristic silence.

The mystery didn't last long. A few days later one

of the staff came to me and warned me to keep an eye
on Mitch, for he had been observed stealing everything
he could get his hands on, from maple syrup and cherry
brandy to gasoline siphoned from our delivery trucks.
The other employees confirmed the allegation, explaining
that because he was in charge of taking inventory he was
in a perfect position to alter the numbers as necessary
to cover his thefts. Gordon and I were stunned. Why
hadn't they told us sooner? It turned out they were afraid
of Mitch, who had the reputation of being remorseless
when wreaking revenge on anyone courageous enough
to tattle on him.

Gordon and I watched and waited. We noticed a good
many items disappearing from the factory, but we were
never able to catch the culprit red-handed. Our man was
evidently very circumspect, but we knew that if we gave
him enough rope he would eventually hang himself.

It wasn't long before the expense account incident
gave us all the proof we needed. We had entrusted Mitch
with a considerable sum of money to cover his costs while
making a special delivery for a promotion in New Bruns-
wick. When he returned from the trip his receipts indicated that he had devoured enough food at his motel to
satisfy two or three hungry men. We had no objections
to his eating all the food he wanted, of course, as long
as it was a bona fide expense, but it looked to us as though
a good portion of his meal receipts had been transformed
into cash and pocketed by Mitch. I wanted to be absolutely positive, however, that my assumptions were correct before making any accusations, so I drove all the
way to the motel where he had been staying and inter-
viewed the manager, the dining room attendant, the
waitress, and the front desk clerk. All agreed that Mitch
had falsified his accounts, and they provided me with the
records to prove it.

This may not have been the most serious crime in

the world; nor were Mitch's methods particularly original, but it told us a good deal about his turn of mind. Disappointed as we were by his dishonesty, we nevertheless agreed to give him another chance if he showed even the slightest remorse or a desire to change his ways. But we were in for a surprise. When we expressed our disappointment in his behavior, he unblushingly denied everything and smoothly tried to blame the personnel at the motel for making a mistake on his bill. It was easy to prove there was no question of a mixup or an oversight, but Mitch felt no compunction whatsoever about making one false statement after the next in his effort to clear himself. What depressed both Gordon and me was the ease with which he switched from one story to another as he tried unsuccessfully to cover his tracks. He was perfectly relaxed as he fielded our embarrassing questions, and he met our gaze with unwavering eye contact. He was obviously an experienced liar, and we were saddened to think he had learned his techniques at such an early age. We replaced him the next day with another young man who we hoped would have more sense than to give up a promising career for such little gain.

· "What I can't figure out is where that lidder could have *gone*," the production supervisor was saying as he wiped his hands on his lab coat. "We looked everywhere for it, and I mean *everywhere*. We've turned this whole place upside down. We even went through the garbage. I just don't understand it. The lidder was such a big piece of equipment to go missing that way!"

As he spoke I could see Mitch's cold and sullen eyes staring me down as I told him he was fired. I remembered the stories I had heard about his vengeful ways, and it didn't seem unlikely to me that Mitch himself was responsible for the mysterious disappearance of the lidder. Even though one of the staff had kindly agreed to stay in the house while we were in Europe, Mitch still could have sneaked into the plant during the night and made off with the lidder. Since it would cost well over a thousand dollars to replace the part, and considering the time required for the manufacturer to machine a new piece for us, it seemed worthwhile to make a determined effort to find out if Mitch did indeed know something about the whereabouts of the missing lidder.

The local branch of the Royal Canadian Mounted Police was most cooperative. They immediately brought Mitch in for questioning, but were unable to draw any conclusions from their initial contact with him. They then proceeded to query everybody else who worked for us, taking great pains to examine the factory to see whether or not it was possible that the part could have been mislaid. When their first round of probes turned up nothing out of the ordinary, they methodically began a second round of questioning.

It was at this point that Mitch panicked. Evidently he must have thought the Mounties were beginning to zero in on him, for he felt it expedient to accuse one of his friends of stealing the lidder. The friend was justifiably outraged by this false accusation, so he immediately told the Mounties that Mitch was indeed the culprit. He had apparently boasted to several buddies about his exploit, and it was by then common knowledge that Mitch had thrown the lidder off the Government Wharf and into the Back Harbour. It wasn't long before the Mounties elicited a confession from Mitch, who was later forced by a court ruling to reimburse us for the part and donate a hundred hours of labor to the community. Nothing, however, could really make up for the many hours of unnecessary labor we and our staff underwent as a consequence of Mitch's abysmally faulty reasoning. He was fired for stealing? Then what could be more logical than to get back at us by stealing again?

Not long after the Mitch incident a man came to us to apply for a position as plant manager, and he told us a horrifying story. Apparently he had once owned a meat-packing plant in Ireland, and he too had fired an unruly youngster for some sort of misbehavior. The boy had then turned around and burned down his plant. Because of difficulties with Irish law enforcement at the time, he was unable to get his insurance company to pay him for his loss, so he and his family were left destitute. He had eventually emigrated to Canada, where he earned his first mate's papers, but his job kept him away from his family for months at a time. The young arsonist in Ireland had managed to affect permanently the life of the man who had been kind enough to give him employment.

Gordon and I had been lucky. The Mounties had gotten their man, and we had escaped relatively un-scathed from a situation that could have turned out much worse. But from that time on I became more acutely aware of the many risks involved in owning a small busi-ness, not the least of which is one's vulnerability to the ploys of the unscrupulous and the mediocre.

The rampant mediocrity factor

Mediocrity, as my irascible father was wont to remind me, comes in many guises and can be counted upon to be rampant everywhere in this vast desert of human imperfection. As a child I would watch him in fascination as he tirelessly elaborated on the subject, his face florid with indignation and his jugular veins bulging with out-rage. At the time I used to think he was rather liberal in his use of hyperbole, but experience taught Gordon and me to adopt an extremely sympathetic view of his opinions. We grew so used to the rampant mediocrity factor that it eventually became a family joke, and soon we began to refer to it by its acronym—RMF.

The RMF dogged us wherever we went, and negligence was not the least of its many faces. Delicate stainless steel parts would be dropped carelessly on tile floors; stacks of yogurt would be overturned during forklift horseplay; and truck engines would burn out when "everyone else" forgot to check the oil. But one of our worst financial disasters hit us in the middle of a blizzard when a warehouse collapsed on top of forty thousand dollars' worth of frozen raspberries belonging to us. The warehouse owners denied liability and forced us to take legal action to recover our loss.

Worse than the financial pain to us, however, were the long-term effects of losing a year's supply of frozen raspberries. After making hysterical phone calls to brokers all over the United States and Canada, we were finally able to locate enough new raspberries to last until the next pick, but the replacements were *whole* frozen berries, which were considerably more expensive than the crumbled fruit we had been using before the blizzard obliterated them. At first we thought the whole berries would make even better yogurt and be greatly appreciated by our customers, but soon the complaints started arriving. The yogurt was now a much lighter color and had fewer pieces of fruit, and how dare we cut back on our recipe? Gordon sat down at the typewriter and wrote dozens of letters explaining that the whole berries didn't thaw as quickly as the crumbles so the juice couldn't disperse itself as easily throughout the yogurt, and although there weren't as many pieces in the product the berries were considerably larger than before.

Once the people understood the situation they were thoroughly mollified, but how many customers noticed the difference and quietly decided to buy some other brand without consulting us? We would have liked to take out a full-page ad in all the major newspapers to explain what had happened, but by the time we had paid for the new supply of raspberries there was nothing left to spend on such extravagances.

It wasn't long after the frozen-food warehouse debacle that we were peering once again over the edge of the abyss, for nothing in our entire history could compare with the mischief caused by the RMF in connection with our filling machine. It arrived safely from Denmark one summer afternoon accompanied by a cheerful Danish mechanic who carefully and lovingly installed it in our plant with the help of a talented local electrician. There it stood, seventy-five thousand dollars' worth of stainless steel and electronics. It was especially designed to be idiot-proof, for we had explained to the manufacturers in Copenhagen that neither Gordon nor I nor anyone on our team had any great mechanical aptitude and that it would be difficult for us to cope with a breakdown. Accordingly, the designers had provided us with a machine equipped with an oil pump that would completely oil every part of the filler. All we had to do was pull the little handle twice a day. That seemed simple enough, but just to be on the safe side we hired a tall, black-eyed, moustachioed mechanical engineer named Mortimer Birdsall to oversee the general maintenance of our equipment. He had several degrees in mechanical engineering and many years of experience as maintenance manager in various plants in Alabama.

Gordon and I sat back and heaved a sigh of relief. At last we were covered on every flank, and nothing disastrous could possibly befall us anymore. Or so we thought. On Mortimer's third day he announced that a part was missing from the automatic oil pump on the new filling machine, but not to worry. He had ordered the part from the Danish representative in Toronto and it would arrive on the next plane from Copenhagen. Meanwhile, he would take the machine apart every evening and oil it by hand. Gordon was delighted that we had a man on board who knew enough about the filling machine to strip it down and keep it running in perfect order.

Five weeks went by and the missing part still hadn't arrived from Denmark, but Mort was not in the least bit concerned. The part was actually made in Germany, he explained, so it was not surprising that the Danes were having a little trouble locating it. These things take time. This explanation sounded normal enough except for one small detail: I was beginning to notice that where Mort was concerned *everything* took time. He had not yet fixed the gears on the custard kettle; and he had neither stopped the leak in the pipe over the ice bank, nor yet found the time to replace the plastic stripping at the back doors of the forty-two-foot trailer. The list went on and on, and as time went by I began to feel more and more nervous. I finally lost my patience when he allowed the refrigeration unit on one of the trailers to run out of fuel. He couldn't seem to bleed the unit, so we had to call in a specialist from Dartmouth. Meanwhile, several thousand dollars' worth of yogurt was sitting unrefrigerated in the afternoon sun. When the mechanic from Dartmouth finally showed up, it took him exactly thirty seconds to bleed the unit and get it going, after which he gave us a bill for two hundred dollars and went merrily on his way.

Mort was defensive when I asked him why he had allowed the refrigeration unit to run out of fuel in the first place.

"I just didn't have time to check it, that's all," he said.

"It only takes five seconds," I persisted.

"Don't worry about it," he snapped. "I'll keep an eye on it from now on."

"What sort of a system are you going to use to remind yourself to keep an eye on it?" I wanted to know.

"I won't need anything to remind me. It's not going to happen again," he said, glaring at me.

It took me only five minutes to sit down at the typewriter and make him up a check schedule for the following two months. I asked him to sign it for me every day, but I noticed that he immediately assigned the task to the cleanup crew. Evidently he didn't have time to do it himself.

It seemed only logical to assume that if Mort didn't have time to do anything tangible, no matter how small the task, he didn't have time to oil the filling machine, either. The next morning I had a little chat with my production supervisor.

"Thank God you came and asked me about it," he said, obviously relieved to see me. "I'm worried to death about the filling machine. It's been vibrating for the last two weeks, and now it's beginning to make funny noises."

"But Mort says he oils it every day!" I exclaimed. "What could possibly be wrong with it?"

"Nobody's seen Mort oil it even *once*," he said firmly. "And there's always somebody around. Anyway, that machine can't be oiled by hand. It has no nipples that I can see. There's no place to insert an oilcan, so he couldn't lubricate it even if he wanted to!"

"So if you were worried about the machine," I said in exasperation, "why didn't you say something to me about it?"

"Mort would have killed me! He's told us all not to go to you about anything. He says he's the plant manager and he's in charge now, and nobody is to go over his head!"

I felt my face flush as the anger mounted. The man had effectively cut off all communication between my staff and me without my even knowing about it, and at the same time he was using normal plant procedure to cover his backside. I got on the phone and within half an

hour I had established a network of referrals and rec-
ommendations made by various maintenance managers
at plants near and far. All fingers seemed to point in the
direction of two highly qualified Lunenburg men who
were known to be excellent at their jobs. One was a ship's
engineer and the other was a maintenance worker in the
frozen fish department at National Sea. Each was willing
to take a quick look at our filling machine and pronounce
judgment on it.

Both men came on separate evenings when Mort was
absent, and both had the same thing to say about the
filling machine. Most of it was dry as a bone and desperately in need of oil. Moreover, some pieces were seized
up and others were rusty because somehow water had
seeped into what little oil had managed to find its way
into the interior. In short, the filling machine was in
critical condition and needed to be overhauled at once
before it turned into a useless piece of junk. We hired
the ship's engineer to do the job, and within three hours
he had personally fashioned a part for the oil pump and
got it working again, and in no time he had removed the
rust and freed the seized parts on the inside. In one day
he had accomplished what Mort could not have done in
any amount of time.

There was little doubt in my mind that the ship's
engineer had rescued us from what might well have been
bankruptcy. It would have been hard enough for us to
find another seventy-five thousand dollars to replace the
filling machine, but the downtime would have lasted at
least four to five months. By that time we would have
lost all the shelf space we had so painstakingly built up
in the stores, and our carefully selected employees would
have long since found other jobs. If we had waited any
longer to get second opinions on the condition of the filling

machine, Mort might have managed to put us completely out of business.

In spite of everything, we felt bad about having to fire him. He naturally thought a great deal of unnecessary fuss had been made over a situation that he had well under control, but we told him firmly we were going to contract the maintenance out to more qualified people. He challenged me to find a more qualified person than himself, and I had to admit that Gordon and I had both been impressed with his résumé. He had every known degree in mechanical engineering, but his knowledge must have been purely theoretical. It was hard for me to understand how he could have lasted as maintenance manager in his other places of work, but I supposed he must have become adept at shuffling papers and generally concealing himself from the scrutiny of his superiors thanks to the sheer magnitude of the companies where he had been employed before. One thing about working for a small business—there's no place to hide.

Our experiences with both Mitch and Mort brought us up short. They taught us we were vulnerable and eternally at risk when it came to people, but we weren't quite sure how to go about minimizing the risks other than to vow never to make the same mistakes twice. From Mitch we learned the advisability of withholding trust until it was earned ten times over. As for Mort, I noticed his hands for the first time when I gave him his separation papers. That day I promised myself I would never again hire a maintenance man with clean fingernails.

We dismantle the house that Daisy built

"Gordon, why don't we sell the cows?" I said one evening. It was eleven o'clock and Gordon was only just getting back from the barn. His overalls reeked of manure and his hair was damp with perspiration.

"Sell the cows?" he echoed. "But we need them to gnaw off the land."

"We can rent the pasture to the neighbors. Travis would be glad to put his critters on our grass. And if not, there's always Cecil Mosher, and Burt Whynot, and plenty of others. They'd be happy to make the hay, too, if we let them haul it away for nothing. They'd probably even make fence in the spring if we gave them free pasturage."

We had had this conversation many times in the past, but it was always difficult to come to terms with the idea of getting rid of the cows. Gordon had fed them and milked them and cleaned them out for so many years that he had learned to take his slavery almost for granted. Besides, we had both developed a feeling of affection for the critters we had nurtured for the better part of a decade. Nevertheless, it seemed best to close this chapter of our lives before the strain took too much of a toll on us.

"I don't know," said Gordon, shaking his head. "There's a statement to be made about farmers, too. I wanted to see if I could find a way to help them make a living if they combined farming and business. I haven't finished experimenting with the idea. Why should it be more profitable to produce Coke than milk? It makes me mad, you know."

"Well, you can think about it all you like, but it's not going to do you any good to kill yourself with overwork. Just think what you could do for Peninsula Farm if you gave it your full attention! As long as I'm working fulltime at Dalhousie there's not much I can do to help except keep an eye on things, but if you had the time you could go out and haul in all kinds of new accounts. We're not in the hospitals, or the schools, or the institutional cafeterias. . . . There's no end to the possibilities! You could do so much more for the company if you were out there selling yogurt rather than mucking out the cows morning and night. Do you realize how much time you spend in that barn? Anyway, it's silly to be milking cows when we're buying most of our milk from Twin Cities now."

It took Gordon well over a month to accept the fact that we would be better off without Daisy, Clover, Bambi, Melody, and the rest of the girls. But the more he reflected on the joys of having his evenings free and spending more time with his family, the easier it was for him to visualize himself leading an essentially cowless existence.

The auction was hard for all of us. We knew we would miss our old friends, especially Daisy the Cow, but we comforted ourselves with the knowledge that her picture would always appear on our plain yogurt containers as well as on our letterhead. Daisy had been trademarked, after all, and we had to admit that this was more than most cows could say for themselves. Gordon, for his part, suffered for about two days from an acute case of postauction blues. But on the third morning he jumped out of bed and announced that he now felt as though he had just been released from ten years on a chain gang, and from that moment on he never looked back.

The departure of the cows triggered a general exodus of barn animals—a sort of balance of nature act in reverse. As soon as the grain disappeared, the mice became disgusted and left in a huff, making life very boring for the cats, who also decided to pack up and move to a more promising barn down the road. Even the maggots found life less fulfilling. Their population declined so much that we ended up getting rid of the chickens, too. As for the pigs, there was no longer sufficient body heat in the barn to keep them warm over the winter, so we decided not to buy any new litters of piglets after the Christmas slaughter. Besides, there were not enough mistakes coming out of the product development department to satisfy their insatiable appetites. My experiments with puddings had come to a halt when I realized they would not last much longer than ten days without spoiling. We thought about canning them or at least packaging them with ultra-high-temperature processing equipment, but we quickly determined that such machinery was well beyond our means. We were very reluctant to consider using preservatives to retard spoilage, but the final blow was the discovery that the puddings could not be trusted to remain stable without the addition of sodium stearoyl-2-lactylate and other chemicals we deemed incompatible with our company's goal as a purveyor of pure food.

Our failure to launch a pudding line did not mean that our search for the perfect chocolate in Holland and Switzerland was to go unrewarded. Although our home-made ice cream has never been a profitable item for us because of the exorbitant cost of the ingredients and the back-killing labor of grinding it out, the public has unilaterally decided that we are to continue making it. Every year I firmly announce that I can no longer subject myself or my employees to the frustration of custards that want to scramble, batter that tries to turn to butter, and ice cream that becomes crystallized or hard-frozen at the slightest hint of temperature shock. No matter. The customers come swarming to the farm anyway, heedless of my pleas that they turn their attention to yogurt instead.

"No way," they chorus. "We can get your yogurt in any store, but we have to come *here* for the ice cream. We brought our elderly mothers all the way from Halifax just for a taste of the Rum and Raisin, and our relatives came from Toronto to try the Irish Coffee Liqueur. And our children have their hearts set on that Lemon Cheesecake ice cream. You can't send them away with nothing at all!"

The vanillin-free Dutch dark chocolate was blended with cherry brandy and brandied cherries to produce our Black Forest ice cream, or mixed with strong coffee for Mocha, or hand-shaved with a cheese grater for a show-case appearance in Mint Chocolate and Orange à la Donna. The Dutch milk chocolate was combined with pure cocoa and condensed milk for our regular Chocolate flavor, which turned out to be an all-time favorite in spite of its simplicity.

Although there were times when I thought I would collapse at the feet of the ice-cream churn in the summer heat, the torture proved to be worthwhile, for we all enjoyed the contact we had with our vacationing customers. The enthusiasm for Peninsula Farm products was such that by the end of the day our heads were spinning a little, but there was keen satisfaction in knowing that our efforts were appreciated.

Of great disappointment to everyone, however, was the disappearance of the farm animals. Gordon's milking act had become very popular with visiting urbanites who dreamed of one day retiring to the country, and the baby calves, piglets, chicks, and kittens were greatly missed by the junior tourists. I knew that I, too, would be sorry not to see Valerie and Vicki from the kitchen window as they rounded up the cows every afternoon and drove them to the barn for milking. They would march over the fields, four and five feet tall respectively, striking terror into the hearts of the hulking cows with their upraised flyswatters. Over the years the two of them had developed an excellent herding routine based on an intimate knowledge of the personality and mind-set of the individual animals. They knew, for example, that Priscilla would always try to escape into the woods unless Valerie guarded the left flank of the group at just the right moment. They were also aware that Dizzy would get confused and meander off in the wrong direction if Vicki didn't watch her carefully, and Lightning could be counted on to be the slowest cow of all if she wasn't constantly threatened from behind.

I was grateful to the critters. They had taught our girls to have confidence in themselves and to exercise their authority gently but effectively. Perhaps most important of all, Valerie and Vicki had both learned never to let down their guard and never to allow themselves to be taken in by appearances, for they knew that under the deceptively placid exterior of their bovine charges dwelled a cow with a devious mind. They attributed this penchant for mischief to the general boredom of their existence.

"After all," said Valerie thoughtfully, "what is there for them to do? They spend the whole day chewing on grass and jumping on each other. Not too exciting. But when they see *us* trudging along ready to take them to the barn they say to themselves 'Oh boy, here come those two bozos with the flyswatters. Let's give them a hard time!' "

"Now they're gone forever," added Vicki sadly. "But I'm not *too* sorry though," she said, suddenly looking more cheerful. "Now at least I'll have some time to *myself*."

She used her time wisely. After enlisting the help of some of her playmates, she went straight to work making chocolate-chip cookies to go with our ice-cream cones. The approbation of the delighted tourists kept her team going at top speed until she finally saved up enough money to buy a color television for her bedroom. Once this goal was achieved, however, she promptly laid off her little fourth-grade helpers and opted for early retirement. Fortunately, they seemed not to mind losing their jobs, for they had not only made more pocket money in Vicki's employ than they had ever dreamed possible, but the public swimming pool was growing more and more tempting as summer progressed.

We were pleased to see that our two girls ended up with more than just monetary gains from their early participation in the work force. When Vicki entered fifth grade, she came home one afternoon glowing with pride at the results of her first arithmetic test of the new term. She had never been good at this subject, but after running her own business for the better part of the summer she had developed an excellent head for figures. *Vicki's Cookies* (Unlimited) had given her the confidence and experience she needed to succeed.

Valerie, on the other hand, was not interested in starting a company of her own. As the older of the two, she had observed more closely than Vicki the long hours, the constant stress, and the nagging fear associated with owning a small business. Gordon and I no doubt spent more time discussing problems and their solutions than emphasizing the fulfillment we derived from the business, so perhaps Valerie had been presented a rather one-sided picture of the advantages and disadvantages of the free enterprise system. Be that as it may, she worked in the factory every weekend for a full year until she had saved enough money to send herself, at the age of thirteen, to a ski camp in Switzerland. She returned home slim, hard, and aglow from the reflected sunlight at the foot of the Matterhorn, and we all knew we had Peninsula Farm to thank for providing her with an unforgettable experience.

Crabs in a basket

I had always heard that nothing succeeds like success, but as time went by I began to realize that people must have been forgetting to quote the second part of the maxim, for I could not help noticing that in fact nothing succeeds like success in arousing the predatory instinct in others.

Our competitors took an understandably dim view of the excellent publicity our yogurt was receiving, but one particular newspaper article aroused them to new levels of passion. "I happen to be a dedicated yogurt buff," stated the writer of the article in question, "and since I travel widely in Canada, the United States, and Europe, I never miss an opportunity to test the local yogurt. I believe it is safe to say that I have familiarized myself with well over sixty different brands. My conclusion: Peninsula Farm beats them all, hands down. It is undeniably a world-class yogurt, and it's made right here in Nova Scotia."

At the next dairy convention, I was accosted by one of our competitors who enthusiastically squeezed my hand almost to the breaking point. "You've been getting some pretty fair publicity lately," he said heartily, flashing a wooden smile and fixing me with a glinty stare. "But that last article was a real snow job. The best of sixty brands! How'd you get the journalist to make *that* kind of a claim, eh?"

He nudged me in the ribs with his elbow and gave me a conspiratorial wink before turning away.

Later that evening I happened to meet his wife. As the man was obviously posing as a yogurt connoisseur by stating categorically that the reporter had given the public a snow job, I was curious to know how many different brands of yogurt *he* had tasted, and what his preferences were.

"Who, *Bernie?*" cried his wife, her high-pitched laughter bubbling over. "You couldn't get Bernie to eat yogurt if you *paid* him! He's never touched the stuff in his *life!*"

Sad as it was to think that yogurt was being manufactured, advertised, and marketed by businessmen whose major concern was profit rather than the quality of the product itself, I was even more disturbed by the turn of events at Dalhousie. The second edition of my Spanish textbook had raised the total adoptions to over one hundred colleges and universities, and it had been given good reviews by professors, who reported that it was enjoying considerable success with their students. As a result of this and other publications, I had risen in a normal fashion through the ranks at Dalhousie until I was finally promoted to full professor with tenure. This must have been more than my supervisor could bear.

"I'm sorry to have to tell you this, Mrs. Jones," he said one day, peering at me over his glasses. "But I'm recommending to the dean that he not grant you your career development increment this year. I feel your career has been distinctly impeded by your involvement in the family business. Why, just the other day I heard somebody refer to you as the 'yogurt queen.' I think you can appreciate how undesirable that is for the reputation of a university such as ours. Anyway, I'm truly sorry to have to take this measure. Believe me, I agonized for many nights over the decision, but I felt a little shock of this sort might help you to concentrate your efforts more firmly on your academic duties and responsibilities."

"I don't understand," I stammered, trying to compose myself. "That increment is withheld only in cases where a professor's performance is clearly substandard. Surely you can't possibly think my work is *that* poor!"

"If you consider for a minute the definition of the word *substandard*, Mrs. Jones, you'll quickly realize that it could apply to exactly half the teaching staff at Dalhousie."

"In theory, yes," I agreed. "But in practice the CDI is withheld only in a few cases a year, where professors are obviously not doing their jobs. It only applies to the bottom one percent, and I don't belong in that category."

"I'm afraid I'm not going to be influenced by what others do," he said smugly. "If they don't have the courage to withhold the CDI in those cases where their colleagues are not performing up to the standard set by the university, then that is *their* problem, not mine. They'll have to live with their own consciences."

"I get up at three o'clock every morning," I protested. "I've given up friends, recreation, social activities, and leisure time. I haven't been cheating the university in any way. I've accomplished as much as anyone else, and probably more. And I'm a good teacher, too. My students will tell you!"

"I'm glad you have such a high opinion of yourself, Mrs. Jones," he said, with a sardonic smile. "But unfortunately you're not the one making the decisions here, and I say that nobody can run a business and be a successful professor at the same time."

"I *told* you," I reminded him. "It's the *hours*. I work a good sixteen hours a day!"

"Let me put it this way," he said, leaning back comfortably in his chair. "I'm not a Communist, you understand, but I do subscribe to one of their dictates. I definitely go along with them when they say 'to each according to his needs, and from each according to his ability.' Now, if you have the ability to work sixteen hours a day, and that is very admirable indeed, then you should be dedicating every one of those sixteen hours to Dalhousie."

It was useless to argue with him. Instead, I was forced to take my case to the grievance committee, who had the information necessary to ascertain that my administrative contributions were far higher than average, and that my publications, if judged on sheer volume, put me well into the top 10 percent of the academic community. If judged on quality, it was hard to argue with the popularity the textbook was enjoying in both Canada and the United States; neither could any fault be found with the reviews that had been written about my book *Alfonsina Storni* by professors in several other universities. As for my teaching ability, the grievance committee managed to come up with an evaluation sponsored by the Student Union. My classes ranked me in the top 5 percent of the university. There appeared to be nothing even remotely substandard in anything having to do with my performance at Dalhousie.

The administration was embarrassed. The recommendation to withhold my CDI was quickly overturned, and the figures on my next paycheck indicated that restitution had been made and all was back to normal again.

I was vindicated, to be sure, but I felt that nothing would ever be quite the same again. It was never my intention to tangle with my supervisor, but I knew that from then on I would have a bitter enemy in him, for he no doubt blamed me for his own humiliation. I also sensed that once mud is slung it tends to stick, unfair as it may be. In years to come would people remember the outcome of the story, or would they recall only that I was involved in some sort of an uproar having to do with my being substandard? It seemed like a sad ending to a decade of conscientious work.

"I know how you feel," said Gordon sympathetically. "But don't let yourself be discouraged. You have options, and you can make choices when the time is right, but most people are like crabs in a basket. I'll always remember one day when my dad took me to the shore one weekend to go fishing. We were standing on a wharf waiting for our boat and watching this old guy catching crabs in a net. I must have been about ten years old. Anyway, the fellow was dumping his crabs into one of those old-fashioned hemp baskets made of coiled rope, you know the kind I mean?"

"I think so."

"Well, I stood there watching those crabs in the basket, and I saw them start to climb up the sides. It was easy for them to claw their way up because the sides were so rough. I thought the old fellow was crazy to use a basket to keep his crabs in, rather than something like a pail or a tub with a smooth surface. So I told him his crabs were going to escape if he didn't watch out. I was worried for him. He looked me up and down for a while without saying anything at first; then he told me to keep an eye on those crabs and let him know what I saw. So I stood there and I watched them all right, but every single time one of the crabs almost made it to the top of the basket, another crab would reach up from below and pull him down. And they kept going like that till our boat finally came and we had to say good-bye to the old fisherman. But I've never forgotten him, or those crabs of his, either. He sat there smoking a corncob pipe with his back turned to the basket. He knew he could count on them to keep one another where they belonged. He never even bothered to glance over his shoulder. I thought those crabs were just about as stupid as they could be, and I can say the same thing for your supervisor, too, and maybe for one or two other people we've met. But don't let them bother you. Just stay out of their basket, that's all. Turn your back on them, like the old fellow on the wharf, and let them play their silly games with each other while *you* do the fishing."

16

Revelation

Toronto or bust

As our upwardly mobile costs continued their relentless climb, Gordon and I kept on struggling to find new ways to improve our production volume. Sales were starting to level off again, so it was decided that while Gordon took care of the day-to-day business at home, I should go forth into the world and find new yogurt markets beyond the borders of the Maritime provinces.

Toronto appeared to be the city of choice, for one of our former drivers lived there and was eager to work for us again. The situation seemed ideal. He was honest, reliable, hard-working, accurate with his figures, and skilled in the art of keeping trucks and refrigeration units operating properly. All that remained was to convince some buyers and retailers to put my product on their shelves.

It would have been easier to reverse Niagara Falls. The buyers in the chains made Mr. Shoemaker look like a cream puff, and by the end of the week I found myself reminiscing about him with a feeling bordering on affection. His counterparts in Toronto were not content with the "plan" I had worked out with the chains in Nova Scotia. They wouldn't even sit down with me unless I was ready to consider paying them a rebate that was downright exorbitant by Maritime standards. Not only that, but I was expected to fork over a hefty sum of money just for the honor of being listed in the stores, and if I was prepared to get over that little hurdle the buyers were then willing to talk about volume discounts, cooperative advertising, earned cost reductions, payments for shelf space, and other even more imaginative schemes. The best offer I had all week came from a tall man with white hair and a tan face. He listened attentively to my discourse on the long-term profitability of excellence and the benefits that would accrue to his chain if he were to carry our undeniably irresistible yogurt.

"Here's the deal," he said tersely, when I had come to the end of my monologue. "I like the idea of stocking a product made in the Maritimes. It's good for our image to be seen as a big chain helping out a little manufacturer from a have-not province. So I'll make it easy for you. All you have to do is pay me fifty thousand dollars up front, and I'll take your yogurt on a test basis for six months. If it sells, I'll list it. If not, we'll call it quits."

"And the fifty thousand?"

"No refunds."

I decided to forget about the chains and concentrate on the small, independent stores in the hope that their lack of power would render them somewhat more accessible to me at my present level of development. But I soon learned that many of the independents had banded together for protection and were supporting one another in their demands for supplier rebates similar to those exacted by the large corporations. The rebates requested even by these modest outlets were well beyond my means.

"I'm really sorry we can't help you," said the owner of one small store. "But we'd have gone out of business long ago if we hadn't joined forces to compete with the chains. So I'm afraid you'll have to come up with the same payments as the other yogurt suppliers here in Toronto if you want to get in."

I discovered only one store owner who had preserved his independence and managed to flourish in spite of the whirlwind of deals that had long since carried everyone else away. He was a crusty old fellow whom I found holed up in the back room of an apparent warehouse hidden in an alley in the downtown area of the city.

"A new yogurt, eh?" he grunted when I introduced myself. "You got a sample?"

I handed him a small container of raspberry yogurt, which he accepted without comment. He turned it around several times, scrutinizing it with a jaundiced gaze before peeling off the lid. He sniffed it skeptically, then reached into his drawer and pulled out a wooden coffee stirrer. He scooped a small amount of yogurt from the container and tentatively smeared it over the tip of his tongue. For a moment his eyelids drooped while his mind concentrated on the flavor and consistency of the substance in his mouth. Then suddenly he opened his eyes wide and sat bolt upright, staring dumbly at the container in his hand. In a flash he tilted his head back and with one movement of the arm he dumped the entire contents of the cup down his gaping throat.

"I must have this yogurt," he said, burping comfortably. "It's like you went out in a field and picked raspberries and dropped them in here. How'd you do it?"

"They're IQF raspberries—instant quick frozen. They freeze each one of the berries individually to preserve the juice and the integrity of the fruit."

"I gotta have it. How much you want for it?"

"Sixty cents a container. But I'm afraid I can't pay

you as much of a rebate as you're used to getting here in Toronto. . . ."

"I don't give a damn about that," he said scornfully. "When I come across something I like, I buy it. I don't worry about anything except whether the product is good or not. You know what your costs are. If you need sixty cents a container, that's what I'll give you."

"How can you be competitive, then?" I asked.

He looked at me curiously for a moment, then broke into a toothy grin. "Low overhead," he chuckled. "Haven't you noticed?"

His store was indeed a study in the concept of the cost-plus warehousing that has become popular in the last decade. Goods were carelessly arranged in random piles and displayed for the most part in their original packing cases. What few clerks there were seemed overworked and underpaid, but the store was packed with customers.

"I always come here to buy my groceries," remarked one shopper. "The prices are reasonable, and I can get things you can't find in the supermarkets. There's lots of locally made food here, and ethnic stuff, and just plain down-to-earth ingredients that are nice to have around the kitchen, you know what I mean? I like to prepare wholesome meals for the kids. I'm a start-from-scratch type of person, and this is my kind of store."

Unfortunately there weren't enough such stores in Toronto to justify shipping our product so far from the plant. Transport trucks are not interested in small loads, and the refrigerated rigs are usually packed with fish— poor bedmates for yogurt. I returned home feeling excited by the prospect of someday penetrating a market as dense and promising as Toronto, but it was obvious to me that we were not quite ready to compete in the "big leagues" yet. Toronto was a vital, bustling city where consumer products moved quickly off the shelves, so it was only natural that the stakes should be higher where the potential for profit was so much greater than in the Maritimes. I knew we would have to wait until Peninsula Farm was financially strong enough to hold its own among the industrial giants in the food business, but it was hard to be patient after personally viewing the multitudes of innocent customers calmly buying their usual quota of dairy items, unaware that on the south shore of faraway Nova Scotia was a plant that produced the yogurt of their dreams.

Tokyo Blues

"There are other alternatives, you know," Gordon mused as we sat at the kitchen table discussing the results of my trip to Toronto. "We don't necessarily have to wait until Peninsula Farm can pay its own way. We could grow faster if we wanted to explore some deals with a venture capitalist, for example."

I looked at him thoughtfully. "They call them 'vulture capitalists,' don't they?"

Gordon smiled knowingly. "They do have that reputation, and maybe they've earned it. But we could be careful. We'd have to explore the terms in detail before committing ourselves to anything. It's an idea, anyway."

"It worries me. I don't trust them."

"We could sell some equity to friends, if that would make you more comfortable."

"It wouldn't, though. How do you think we'd feel if things didn't work out the way we expect? I'd hate to be responsible for making a friend lose money."

"Well then, we'll just have to grow slowly," Gordon sighed. "It's probably the sensible way to go, but it drives me up the wall."

"If it works, don't fix it. Isn't that what your friends in New York used to say?"

"Yes, and they were right. What we're doing is working pretty well for us. It's just that I get impatient sometimes, but I'll live with it. The thing to remember is that quality is everything. First, last, and always. If we keep up the quality our reputation will gradually spread, and eventually the chains will be inviting us in just to satisfy their customers. If this happens the rebates won't be so important anymore. Smart buyers are always sensitive to consumer demand, so if they want our product they'll take it on our terms. We'll succeed all right, as long as we keep on making outstanding yogurt."

It was for this reason, among others, that we delayed the launch of our wild blueberry yogurt for two interminable years while we worked to find a way to make the obstinate blueberries behave themselves in their yogurt medium. Just when we were on the verge of despair, Rick Lawrence of the Agriculture Canada station in Kentville came up with a two-part answer to the problem. The first part of the solution had to do with a recipe whose success depended on the accurate control of time, temperature, and sugar content, for the sugar had to coat the blueberry skins just enough to prevent the juice from flowing out freely, but not enough to make the skins crystally and chewy. This was a great eye-opener to me, for the sugar content was the one element I had neglected to control in my previous experiments.

The second part of the solution came as a complete surprise, for Rick was quick to point out that we had the Japanese to thank for the success of the new recipe. In recent years, he explained, the Japanese had gone wild over blueberries, and they were buying them from Nova Scotia by the carload. But their standards were extremely high, and they refused shipment on any berries that were not up to their expectations in size, freshness,

ripeness, and appearance. There were only two growers
in Nova Scotia, as far as Rick knew, who could meet the
Japanese standards, and they packed special blueberries
known as Tokyo Blues. Even though the Tokyo Blues
were unusually expensive, Gordon and I decided to use
them in our wild blueberry yogurt because of their exceptional flavor and tenderness. Our customers were obviously delighted with the carefully selected berries, for
no sooner was the wild blueberry yogurt on the market
than it outstripped even our best-selling raspberry flavor. We took our hats off to the Japanese for having
created the demand that led to the development of Tokyo
Blues.

"The blueberry launch was a great success," said
Gordon at the breakfast table one morning. "But I'm
afraid we can't afford to rest on our laurels. There are
still piles of bills to pay. So now that I'm not tied to my
milk stool anymore, I'm going out to see what I can do
about the institutional market."

The provincial government was putting on a campaign to encourage any institution enjoying public fund-
ing to buy local products, so it wasn't long before schools,
hospitals, and convention centers were calling in their
orders. It was not always easy, however. One university
residence declined our yogurt because it was too good.
The students were paying a fixed price for their meals,
so the manager was not eager to provide them with an
item that was so delicious that they might be tempted
to go back for second helpings.

Other institutions found ingenious excuses for not
taking our yogurt. One claimed that the patients in their
hospital were refusing to eat our strawberry yogurt on
the grounds that it was too brightly colored and must
therefore be artificial. When Gordon explained that our
product was all natural and contained more fruit than

any other on the market, they then complained that it was *too* fruity.

Another government-funded institution declined to buy our yogurt because they only wanted product packed in 125-gram containers, and ours, unfortunately, was packed in 175-gram tubs. Gordon measured the difference and found that it amounted to only three tablespoonfuls, but the institution remained adamant. There are only two yogurt producers in the Atlantic provinces, ourselves and Twin Cities Dairy, and we both pack our small sizes in 175-gram tubs. It wasn't until we pointed out to the buyer that he was therefore restricting his purchases to out-of-province manufacturers that he finally relented and sent a purchase order over to Twin Cities.

Many other institutions claimed it was too much trouble to buy yogurt from us and milk from another dairy, and still others asserted that they were locked into binding contracts with Quebec dairies, but Gordon persevered and slowly won them over. Our volume increased dramatically over the following months, and everyone was pleased to see our new equipment being used to capacity.

Gordon also contributed to our steadily increasing sales by doing in-store demonstrations on a sporadic basis whenever he had the time. He never ceased to be amazed at the number of customers who *still* had not heard of Peninsula Farm yogurt, in spite of our nine years on the supermarket shelves, but he was equally gratified by their positive response when they tasted it for the first time. Not everyone, however, was willing to give it a try. A number of shoppers had become cynical about salesmen claiming to have products that were special, or outstanding, or even just plain wholesome. And many others harbored the old prejudices about how sickening and disgusting yogurt was supposed to be. But 99 percent of Gordon's experiences with the customers gave him great satisfaction. Those who were already Peninsula Farm yogurt fans were delighted to meet him, and they often recounted stories about themselves or people they knew who had never been able to eat yogurt until they had tasted ours. Gordon would come home tired but happy, saying he had met literally hundreds of customers who made him feel our years of travail had not been in vain. I had vowed long before never to compromise our quality in any way, but it was at times like those that I realized our very survival depended on my loyalty to that promise.

Gotcha!

It has been said that success in business is often predicated on an individual's ability to drive himself into the ground, to hang on like a bulldog, and to dream the impossible dream no matter what the odds. Few realize, however, the important role that plain, old-fashioned luck plays in this agonizing process. The pathway to success is so strewn with failure that every small triumph along the way is seen by the struggling entrepreneur as a credit to his ingenuity, creativity, and general tactical brilliance. But there are times when luck is undeniably a decisive factor in the growth of any small business.

This fact was brought home to me when Brian Ives appeared at our door. Brian is a cheerful, unassuming native of Pictou County who has dedicated his attention in recent years to raising, processing, and marketing salmon, trout, and herring. He has a fish farm and smokehouse in Yarmouth, where he puts out some of the province's finest lox. Brian's problem was wastage. He was losing money on the cuttings and trimmings from his valuable smoked salmon, which were often simply being discarded. He had decided to manufacture a smoked salmon pâté, but it was necessary to find a binding medium to hold together the tiny pieces of minced fish. Cream cheese was the obvious choice, but months of experimentation with every imaginable variety on the market had yielded a series of rather dubious results. The major difficulty arose from the various gums and other additives that are routinely used to thicken cream cheese. Brian found them so difficult to handle, in fact, that he eventually decided to abandon the project.

Then one morning at the breakfast table he noticed his wife happily spreading a distinctly ungummy substance on her toast. As he watched her dip her knife into the container for more, it occurred to him that this white stuff, whatever it was, seemed to have all the qualities he was looking for in a binding medium for his minced salmon.

"This? Oh, it's something called 'yogurt cheese,' " said his wife, in response to his inquiry. "It's put out by Peninsula Farm. I don't know what it is exactly, but it's delicious. I'm hooked on it."

It was some time after this fortuitous discovery that Brian came to our farm bearing samples of his new smoked salmon pâté. The product was simplicity itself: minced smoked salmon, yogurt cheese, and aquatic herbs. And it tasted superb. It was so good it was all I could do to salvage some of it from the clutches of our employees, who had proceeded to consume it with the greatest enthusiasm. I did manage to capture one container, however, which I mixed with an equal quantity of yogurt cheese containing thyme and onions. The result was a salmon yogurt dip that disappeared almost as fast as I had made it.

"I'd love to package the salmon yogurt dip myself and add it to my product line," I said wistfully. "But I'm sure you wouldn't want me to compete with your pâté."

"Go right ahead," said Brian heartily. "Feel free. Be my guest. I don't market my products in the Maritimes anyway. There'd be no conflict of interest at all!"

"No wonder I've never heard of your company then," said Gordon. "I was wondering why I'd never seen any of your products in the local stores."

"I couldn't possibly afford to do business in the Maritimes, I'm afraid. There's not enough market here. Not enough population. You have to drive miles and miles just to get from one little town to another."

"Tell me about it," I said, thinking unhappily of our exorbitant trucking expenses.

"No, I send my smoked fish to all the major cities in Canada, and down to the States, too. That's the biggest market of all, of course. Places like San Francisco, Los Angeles, Houston, Miami, Boston, New York. . . . That's where you make the big bucks."

"And what about the salmon pâté? Are you planning to send the pâté down there too?" I asked, feeling the excitement begin to mount.

"Well, that's what I came here for. I wanted to ask you if you were in a position to provide me with large quantities of yogurt cheese. I've already sent samples down to all my brokers, and every single one of them wants to carry the product. They've consulted the distributors too, and they're incredibly excited about it. I think we're going to do well. Maybe *very* well. Will you be able to handle it?"

"You got the product approved?" said Gordon in undisguised amazement. "It has the blessing of the FDA and all the other branches of the U.S. government?"

"Sure," said Brian. "It's all in place. No problem. I did all the legwork before I came here to talk to you. As far as I'm concerned, it's green lights all the way down the road. The rest is up to you. So what do you say?"

Gordon and I looked at each other and burst out

laughing, while Brian watched in amusement. We quickly explained to him how hard we had fought to get permission to send our yogurt cheese down to the States ourselves, and how we had been caught in a catch-22 situation where the uniqueness of our product qualified us to fill a two and a half million–pound quota, which was subsequently denied to us because packaging regulations prevented us from stating the ingredient that made the cheese unique.

"So you see," said Gordon, as he finished the story, "it's really ironic that we'll be sending the yogurt cheese down after all as part of your salmon pâté instead of as a product by itself."

"What's even more ironic," Brian pointed out, "is that now you're going to end up exporting far more than just two and a half million pounds of cheese. If my fish sales are any indication of how this is going to grow, then I think you'd better get yourselves tooled up for a pretty big operation!"

As a final stroke of good luck we were visited the very next week by Leonard Sarsfield, a man well known in the province for having built his wife's recipes for homemade pies into a thriving business. His frozen pies were distributed across the nation and were greatly enjoyed by happy customers everywhere, but Leonard thought it was time to add some variety to the line. Would we be interested in developing a cheesecake with our yogurt cheese? He didn't have to ask twice. The next day I was hard at work testing cheesecake recipes for the Sarsfield operation, and the following month the answer came through: Leonard and his wife Frances would be delighted to manufacture and distribute Peninsula Farm cheesecakes as soon as we could provide them with yogurt cheese on a regular basis and in sufficient quantities to satisfy Canadians from Newfoundland to British Columbia. Gordon and I exulted in the knowledge that at last our yogurt cheese would be granted the recognition it deserved. Here is the recipe that emerged from our kitchen-based department of product development:

YOGURT CHEESECAKE

Serves 8 to 10

Crust:
1¾ cups graham cracker crumbs
2 tablespoons sugar
½ teaspoon cinnamon
½ cup melted butter

Filling:
2 8-ounce packages cream cheese, softened
1 cup sugar or honey
2 teaspoons vanilla
¼ teaspoon salt
2 cups plain yogurt
¼ cup cold water
1 package unflavored gelatin

Preheat oven to 350° F. The cream cheese should be removed from the refrigerator to soften at room temperature at least 1 hour before using.

Combine in a bowl the crumbs, sugar, and cinnamon. Stir in melted butter. Press mixture on sides and bottom of a 9-inch pie pan.

Bake for 10 minutes. Cool and then chill thoroughly in the refrigerator.

Cream the cheese, sugar (or honey, if desired), vanilla, and salt together; then add yogurt and mix until smooth.

Place the water in a small saucepan; sprinkle the gelatin onto surface and warm over low heat, stirring constantly until the gelatin has dissolved. The water will be tepid.

Stir together with yogurt mixture, blending well by hand. Fold into the cooled pie crust, and place in refrigerator overnight.

Gordon realized he would have to get right to work examining the cost and design of the various cheese machines available on the market, for it was clear we would never be able to go on hanging the yogurt in bags if our volume were to increase substantially. It pleased us that we would be including smoked salmon in our business interests, for Peninsula Farm was already supporting dairy farmers, fruit growers, and maple syrup producers. Now we could quite accurately say we were involved in every one of Nova Scotia's major food industries. It made us feel as if we were somehow a part of the growing economy of our adoptive province, and our contribution, no matter how small, gave us a sense of being bona fide members of the community. Peninsula Farm had come of age.

"It's true we've come a long way in a short time," Gordon said, smiling. "Nobody ever thought much about yogurt in Nova Scotia when we first came here. Twin Cities was making a little, but not many people were eating it back then. Now the dairies are pushing it like mad, so we've really started something. I know yogurt consumption is going up everywhere, but three years ago yogurt sales in the Maritimes increased out of all proportion to the rest of the nation. That was right after all those demos, so I'm sure we must have had something to do with it. It's probably safe to say that we got yogurt moving in Nova Scotia, and now that we're in the rest of the Atlantic provinces sales are picking up all over. That tells you something, doesn't it?"

"It tells me that we managed to succeed in spite of ourselves. It's funny when you stop to think about it. We got dragged into this kicking and screaming all the way. If you had told me when I came home that day waving my Harvard Ph.D. that I was going to become a yogurt mogul, I'd have had you committed!"

"Chalk it all up to Daisy the Cow."

"Yes, poor Daisy. She's getting famous, though. Did I tell you my friend Don Patton is beginning to talk about her in his business classes at Dalhousie? All the business school texts emphasize how important it is to *plan* everything, but Don tells his students that growth is often best achieved by adapting, and compromising, and adjusting to changing circumstances, and solving crises. He likes to cite the example of how a yogurt industry was launched by our response to the problem of what to do with the excess milk of one lone cow."

"Well, he's right," said Gordon. "Planning has its limitations. It tends to be sort of bloodless, you know. I'll take someone from the school of vigorous muddling anytime. At least our sketchy, flexible, constantly changing plans have left plenty of room for adventure and humor."

"Maybe so," I said, glancing out the window at our busy little dairy plant. "But we also lost plenty of time and money by learning as we went. We probably reinvented several more wheels than were strictly necessary. And what about your dreams of sailing around the world? We still don't have a yacht, or a boathouse, or even a dock!"

"Ah yes, but don't forget," said Gordon, putting his arm around me as we looked out over the green meadows at the sun-splashed ocean beyond. "Don't forget we recently acquired a very fine rowboat!"

Epilogue

Life at Peninsula Farm continues to provide both Gordon and me with ample opportunity to play a multiplicity of roles as we muddle along trying to meet new challenges, solve problems, and deal with the usual daily emergencies. The general cast of characters remains much the same, with Eddie Shoemaker starring as the villain. It often seems to us that he has a bug planted somewhere in our house, for no sooner do we gain a small profit than suddenly he is on the phone, inviting us to come to his office to discuss how we might increase our rebates to his company.

In fact, his rebate mania has become even more burdensome to us since he attended an international business congress in Europe where he was introduced to the latest developments in the science of improving the almighty bottom line. He has learned all about a total systems approach (TSA) that will allow him to review the direct product profitability (DPP) of his merchandise by analyzing the overall costs on an item-by-item basis. The diabolical Mr. Shoemaker has now fed information about Peninsula Farm yogurt into his new computer program and has concluded that it is high time we increased our contribution to Goliath's bottom line.

As far as I can determine, his conclusions have never once taken into consideration the simple fact that the emphasis on product profitability might possibly contribute to a heightened customer frustration and irritability quotient (CFIQ). What if a shopper walked into a Goliath store and were unable to find the brand of her choice because it failed to meet the new DPP test? I have tried to stress to Mr. Shoemaker the importance of striking a balance between the profitability of a product and the desirability of that product from the consumer's point of view. I feel certain that the grocery items yielding the most favorable bottom lines are not necessarily those most appreciated by the customers. In fact, I am almost tempted to go so far as to say that the quality of a product is in inverse proportion to its profit margin, for the use of cheap ingredients must surely be one of the easiest ways for manufacturers to make a large enough spread to pay an attractive discount to buyers.

But as usual, Mr. Shoemaker is thoroughly unimpressed by my arguments. It would appear that the current trend in the food industry to emphasize product profitability measurements is bound to have a deleterious effect on quality. It remains for Gordon and me to come up with a way to grow big enough to compete favorably with dairies that can afford to make attractive deals and yet stay small enough to preserve our quality by giving our products the personal attention they deserve. The most logical solution, it seems to us, would be to license other dairies to make Peninsula Farm yogurt under our label, using our recipes and procedures.

No sooner did we begin mulling over this idea together than we received a letter from an Irish dairy expressing an interest in discussing a licensing arrangement with us. It turns out that the chief executive officer of the dairy came across our products during a business trip to Nova Scotia and promptly decided that his countrymen would be much better off if Peninsula Farm yogurt were available in their stores. We are currently pursuing serious negotiations with the Irish dairy, and if all goes well we shall soon be doing business in the Emerald Isle. Nothing surprises me anymore.

We continue to owe our real progress, however, to a growing cadre of increasingly savvy consumers. We have found that intelligent shoppers everywhere are becoming more quality conscious all the time, and they seem to appreciate the fact that we put our dollars into the product itself instead of into advertising hype. We believe it is for this reason that we have achieved a phenomenal market share, particularly in stores near universities and professional centers. This indicates to us that we can be optimistic about the way we have chosen to market our products, namely through word of mouth and taste testing.

Our disappointment with the quality of most mass-produced retail food items remains. It is clear that the food-processing giants continue to develop their products in the laboratories instead of in the kitchen. Obviously profit, shelf stability, and other corporate considerations take precedence over the desire of discerning consumers to find top-quality food in the marketplace. If nothing else, we hope this book will inspire consumers, both individually and collectively, to get tough with the food moguls.

Meanwhile, down on the farm, we continue to toil. Now that the critters have found new homes, however, we see less of Travis Oickle than we would like. Once in a while we'll spend some time leaning on our common fence chatting about the price of short-feed, and just the other day he phoned to ask if we were planning to go to the Lunenburg volunteer firemen's fund-raiser, but I had to tell him we couldn't make it. "That's too bad," he said, sounding disappointed. "I wanted a wall switcher." I wasn't quite sure what he meant by a wall switcher, but I told him I would ask Gordon if he could help him out. There was a long pause on the other end of the line while Travis digested what to him must have sounded like a pretty peculiar suggestion. It was Gordon who finally set me straight when he was kind enough to explain to me, once he stopped laughing, that what poor Travis had been asking for was a "waltz with you."

Valerie and Vicki are a little less actively engaged in the day-to-day operation of the business now that homework and school-related activities have begun to make inroads on their time. Once in a while Gordon and I feel a bit nostalgic for the halcyon days when we all worked together as a family team, but we understand that our children must be free to develop their own interests without too much interference from us. Like most normal, healthy kids, they don't spend much time listening to parental advice anyhow. But we know they will find their way; for as we watch them explore the world around them, we realize that both girls seem to have somehow inherited their grandfather's desire to reach for the stars.

Finally, it seems appropriate to end this epilogue with a word about Peninsula Farm's new plant manager. Bill Towndrow was previously employed by an oil drilling company and was responsible for stability and ballast control on a nearby offshore rig. Recently he was granted his master mariner's certificate, which authorizes him to take any ship anywhere in the world. So although Gordon still doesn't have the yacht of his dreams, he at least has a ship's captain to help us steer a steady course through all the perils and pitfalls of the next chapters of our constantly unfolding saga.